The Imagineering Process

Using the Disney Theme Park Design
Process to Bring Your Creative Ideas to Life

THE IMAGINEERING TOOLBOX
BOOK TWO

Louis J. Prosperi

Foreword by Jason Surrell
Creative Director, NBCUniversal

Theme Park Press
The Happiest Books on Earth
www.ThemeParkPress.com

Praise for The Imagineering Process

"Everything is hard if you don't understand it. Everything is hard before it is easy. If you want to understand how Disney Imagineering is able to create such wonderful one-of-a-kind parks around the world, then study this book. *The Imagineering Process* will show you how to use Disney's creative processes to create magic in your own organization. Without creativity your business will be ordinary and quickly become irrelevant. It's never too late to get better!"

—Lee Cockerell, former executive vice president, Walt Disney World Resort, and author of *Creating Magic, The Customer Rules, Time Management Magic,* and *Career Magic*

"Imagineering is the creative scheme that brings Disney theme parks to life. In *The Imagineering Process*, Lou Prosperi unlocks the magic toolbox so you can bring your own creative ideas to vivid reality with all the charm and magic of a trip to Disney World."

—Daniel Pink, author of *When* and *Drive*

"Creativity is merely the arrangement of existing forms into new relationships. Lou has taken a complex creative process, Disney Imagineering, and has broken it down into easy to understand and transferable rules that will guide anybody who is trying to do something wonderful. This book is sure to become the "go-to" book for themed attraction designers and those interested in the art form. In fact, the book is an instructive guide for anybody who is seeking a creative methodology to overcome any obstacle or opportunity. Yes, Lou has done it again."

—Sam Gennawey, author of *Walt Disney and the Promise of Progress City, The Disneyland Story,* and *Universal versus Disney*

"Lou Prosperi lifts the veil on the art and science of creating magic the Disney way! In the experience economy, no business leader can be without this elegant roadmap for winning the hearts and minds of customers."

—Matthew E. May, author of *The Elegant Solution*

"With *The Imagineering Process*, Lou Prosperi does a great job of breaking down the Imagineering creative process...or ANY creative process for that matter. Well-researched and filled with wonderful insight and a lot of passion, this book should be a "must read" for anyone looking to expand their understanding of how Disney Imagineers create some of the most iconic theme park attractions in the world!"

—Brian Collins, former Imagineer & WDI Show Writer

"Ever wonder how to bring the magic of Disney creativity to your everyday work? *The Imagineering Process* translates Disney's seemingly mysterious creative process into simple, applicable insights you can use on any project."

—David Burkus, author of *Friend of a Friend, Under New Management*, and *The Myths of Creativity*

"*The Imagineering Process* points the way toward the elusive meeting point of two crucial needs: the need for new ideas, and the need to produce real things in the real world. Past that, it tethers its thinking to some of the most successful manifestations of creativity and production ever—the Disney theme parks—to show that these ideas work. If you make things, or want to, this is a book for you."

—Jeff Tidball, COO, Atlas Games, co-founder of Gameplaywright, and Origins Award-winning game designer

"*The Imagineering Process* is the perfect companion to Lou's previous book, *The Imagineering Pyramid*—where that book left off, *The Imagineering Process* continues. As a retired Imagineer I can tell you that through much research, Lou has managed to illustrate in a concise, easy-to-understand manner the process that Imagineers use to bring ideas to life, and to correlate that process to other disciplines beyond the berm of the Disney parks. Once you understand this process you will be able to implement it into your personal work with success."

—Louis L. Lemoine, retired Walt Disney Imagineer and Disney Legacy Award recipient

"*The Imagineering Process* is a welcome addition to the study curriculum of all lifelong students of "themed entertainment." In this book, Lou Prosperi shines a spotlight on the importance of sincere collaboration and honest teamwork in the creative process. The greatest resorts, parks, attractions and shows in the world have all come about by adhering to this process in some way, shape, or form. Creating happiness for the world is not easy, but it is worth it."

—Josh Shipley, Chief Creative Officer, Evermore Park and 21+ year Walt Disney Imagineer

"Thank you, Lou, for writing this book to help us crack the Disney Imagineering code. What makes Imagineering one of the powerful divisions of the Walt Disney Company is that they don't just have ideas, they make ideas real. If you've ever had big ideas that you wanted to make real, the process in this book will help you have people standing in line for your creations. It's time that those that create remarkable theme parks don't get to have all of the fun."

—Terry Weaver, speaker, consultant, and author of *Making Elephants Fly: Getting Your Dreams Off the Ground*

To Sheri, Nathan, and Samantha:
You are the three best things in my life.
I love you all more than words can express.
I wouldn't be who I am and couldn't do what I do without you.
I can't wait until the next time we ride the PeopleMover together.

Editor: Bob McLain
Layout: Artisanal Text

ISBN 978-1-68390-138-9
Printed in the United States of America

Theme Park Press | www.ThemeParkPress.com
Address queries to bob@themeparkpress.com

Contents

Foreword

When I first read *The Imagineering Pyramid* by Lou Prosperi, I was struck anew by the realization that many of us are now leading themed lives in one form or another. This notion first occurred to me in 1991, shortly after I moved from Cleveland to Florida to pursue a career in the entertainment industry. *Time* magazine did a cover story on Orlando, which inexplicably featured a picture of the "California Crazy" architecture of Dinosaur Gertie's ice cream stand at Disney-MGM Studios—I could think of plenty of other images that would have better supported their thesis, but at least the art of themed design was getting some national attention. The point of the story was that many of the longstanding principles of themed entertainment were starting to make their way out of the theme parks themselves and into the so-called "real world."

Orlando was understandably ground zero for this movement that was not yet a movement. In addition to the 20-year old Walt Disney World and nearby Sea World, Universal Studios Florida had just opened its gates the year before, sending Orlando well on its way to becoming the unofficial theme park capital of the world. The article pointed out how theme park design flourishes could now be found outside the berm, so to speak, in the form of lushly landscaped roadways, carefully manicured gardens, picture-perfect manmade lakes capped with elaborate water features—even streetlights that suddenly possessed a lot more "character."

Beyond these little design touches that elevated and enriched every civic environment in which they appeared, themed nighttime entertainment districts like Church Street Station and a new contender called Pleasure Island were transporting guests to another time and place for a few hours, providing a form of themed entertainment that had been previously confined to gated attractions. Planet Hollywood's debut later that year, combined with the existing Hard Rock Café, launched the themed restaurant craze that would continue for the rest of the 1990s, resulting in popular chains such as Rainforest Café, Jimmy Buffet's Margaritaville, and Bubba Gump's. Heretofore "ordinary" places were becoming extraordinary, bringing small doses of theme park idealism and escapism into our everyday lives.

Perhaps the pinnacle of this movement, at least in central Florida, was the creation of the town of Celebration just outside the gates of Walt Disney World. Celebration replaced Walt Disney's bold vision of EPCOT with a conscious return to the simplicity and innocence of the past, opting for small-town America over a city of the future. In a way Celebration was still very much infused with the spirit of Walt Disney; it's just that its residents would now be living on Main Street, U.S.A. instead of Tomorrowland.

Everything from the center of town to the residences themselves were seemingly envisioned by a motion picture production designer, which actually made Celebration feel a lot more like a studio backlot than a real town in which real people lived and worked. Yet people flocked there from all over the world, most of them desperately seeking the idyllic existence they knew and loved from the theme parks. They wanted more of that Main Street idyll—all day, every day, if possible. They truly wanted to *live* the dream.

In the decades to come the principles of themed design would manifest in many different areas of our daily lives: planned communities, shopping centers, urban renewal projects—you name it and it was getting an experiential redesign or an aesthetic facelift to elevate the "guest experience" from the mundane to the sublime. And the magical makeovers were no longer confined to central Florida. David Caruso's Grove and Americana complexes in Los Angeles are what you would get had Walt Disney himself ever set his sights on traditional shopping, combining dining, retail, and residential into compelling mixed-use destinations that transported its various constituents to other, better places. On the opposite coast, Disney itself helped spur a successful and equally controversial transformation of Times Square from a decaying urban waste-land into a family-friendly fantasyland in the heart of the city. Granted, I've spoken to more than a few New Yorkers who prefer the old wasteland, but everyone's idea of ideal is different.

In the early 2000s I contributed essays and exercises to *The Imagineering Way* and *The Imagineering Workout*, two books that not only encouraged people to infuse their daily lives with magic, but also gave them pointers on how to do so. Those books were one of my first indications that the principles of themed design were starting to transition from commercial enterprises to personal applications, both in the home and at the workplace. This multi-decade evolution has now taken us from the theme park to the real world to our personal and professional *lives*.

This creative migration has been one I've followed with great interest, and so I was delighted to see Lou key into this movement and demonstrate how the principles of themed design can be applied to our everyday endeavors and help make the mundane magical, the ordinary extraordinary, and help enrich the stories of our lives. And now he's taken an even deeper dive with *The Imagineering Process*, and I'm honored that he asked me to pen a foreword to a book that explores a creative evolution that has done so much to enrich the lives of so many.

The Walt Disney Company, NBCUniversal, and countless other owners and design firms will continue to do what they do so well in the theme park and resort space, creating unparalleled immersion and escapism for their guests. And a small army of restaurateurs, retailers, and real estate developers will continue to apply similar principles to their various real-world endeavors, creating little pockets of magic to help enhance our everyday lives. But now this movement has reached the indi-vidual—it has reached *you*—and I hope you will join it because a little creativity can help make your life better, brighter, and more *fun*. Why settle for the ordinary when the extraordinary is only a little dreaming and doing away?

So I hope you will read this book and come away from it energized, inspired, and motivated, because after all, if you can dream it—well, you know the rest.

Jason Surrell
Creative Director, NBCUniversal

About the Imagineering Toolbox Series

This book is the second in the Imagineering Toolbox series.

The Imagineering Toolbox is a collection of tools inspired by Walt Disney Imagineering, the division of the Walt Disney Company responsible for the design and development of Disney theme parks, resorts, cruise ships, shopping areas, and other venues and attractions.

The *Imagineering Process* is a simplified version of the process Walt Disney Imagineering uses to design Disney theme parks, and can serve as a model for designing and creating nearly any type of creative project.

The first book in the series focused on the *Imagineering Pyramid*, a set of 15 practices and principles used by Walt Disney Imagineering in the design and construction of Disney theme parks.

Introduction

Over the last several years, creativity has become a buzz word in business. A Google search for "creativity in business" returns hundreds of thousands of hits. In his *Creativity Works* column on January 6, 2017, William Childs tells us that his "search on 'books on creativity' revealed that there are over 1.8 million of them." If you read business books, blogs, or websites, you can't help reading about the value being placed on creativity in the modern workplace. In a blog post called "Creativity Creep" from September 2, 2014, on *The New Yorker* website, Joshua Rothman describes this phenomena when he writes:

> Every culture elects some central virtues, and creativity is one of ours. In fact, right now, we're living through a creativity boom. Few qualities are more sought after, few skills more envied. Everyone wants to be more creative—how else, we think, can we become fully realized people?

Simply put, creativity has never been more in demand. In his *Creativity Works* column on January 3, 2017, Childs notes:

> We now find ourselves experiencing what is known as the creative age. It's the age of new ideas, new processes and innovative thought as a way to drive change and solve the complex challenges we face.

In the same article, entitled "The New Age of Innovation," Childs writes:

> While creativity is now being taken seriously as an economic driver, more work needs to be done before creativity is given the full status it deserves. The old stereotype of creative people sitting around on their beanbag chairs looking for meaning through the incandescent glow of their Lava Lamps still lingers.

Quoting Richard Florida, author of *The Rise of the Creative Class*, Childs adds:

> If you are a scientist or engineer, an architect or designer, a writer, artist, or musician, or if your creativity is a key factor

in your work in business, education, health care, law, or some other profession, you are a member of the new creative class.

I agree. I believe everyone is creative, even the people who tell you that they "don't have a creative bone in their body". In their book *Creative Confidence: Unleashing the Creative Potential Within Us All*, authors Tom Kelley and David Kelly refer to the idea that creativity is something that applies only to some people as "'the creativity myth.' It is a myth that far too many people share." They also tell us:

> Creativity is much broader and more universal than what people typically consider the "artistic" fields. We think of creativity as using your imagination to create something new in the world. Creativity comes into play wherever you have the opportunity to generate new ideas, solutions, or approaches.

In *The Creative Habit: Learn It and Use It for Life*, renowned choreographer Twyla Tharp writes:

> Creativity is not just for artists. It's for business people looking for a new way to close a sale; it's for engineers trying to solve a problem; it's for parents who want their children to see the world in more than one way.

So now you might be saying, "Okay, I agree. Everyone is (or least can be) creative. That still doesn't help me actually be creative. How do I do it?" Good question.

I think for many of us the challenge lies in finding the right model of how creativity and the creative process work so we can apply it in our own fields. This book is my attempt at providing just such a model. But before we get to that, I want to briefly look at something that lies at the heart of creativity and that plays a major role in the creative process: ideas.

I believe ideas hold a unique place in regard to creativity. Ideas are at the same time the *most important* and the *least important* part of any creative project. I know, that seems like a paradox, but bear with me.

Ideas are the *most important* part because every creative project starts with an idea. Good ideas are the basis for all successful creative projects. Consider the following:

- Without the idea to create "a place where adults and children can have fun together," there would be no Disneyland (or other Disney parks for that matter).

- Without the idea to develop a technology to allow the creation of human-like robots in theme park attractions (Audio-Animatronics), we wouldn't have attractions such as Great Moments with Mr. Lincoln, the Carousel of Progress, Pirates of the Caribbean, the Haunted Mansion, or countless others.

Of course, good creative ideas aren't limited to those related to Disney parks:

- Without the idea to design and build a separate ship specifically for the moon landing, the Apollo program might not ever have succeeded in landing a man on the moon.
- Without the idea to create a Star Wars-based game for my son's birthday party, my wife and I would have had to entertain 10 young boys all on our own.

Many people (myself included) believe in the importance of ideas, and there is no end to the books, blogs, and websites that offer tips and techniques to help us "be more creative" or "generate new ideas." But generating ideas is not all there is to creativity. It's important, to be sure, but it's only one aspect of the challenge of employing our creativity. Ideas are only a part of being creative, and in some ways (here comes the paradox) they are the *least* important part. What's equally (or perhaps more) important is how we follow through and develop our creative ideas, or as expressed by Guy Kawasaki in his November 4, 2004, Forbes article, "Ideas are easy. Implementation is hard."

If you talk to people in traditionally "creative" fields (writers, artists, designers, etc.), ideas are never an issue for them. Most have more ideas than they could possibly implement in their lifetimes. Generating ideas is the easy part; it's the execution of those ideas that's difficult. The real work is in taking ideas and bringing them to life. Even the best ideas in the world can't execute themselves, and without someone to execute them, even the best ideas in the world have little chance of becoming a reality.

I said earlier that I believe the challenge for many of us lies in finding a model for the creative process—an example that we can look to for concepts and principles that can be applied across a variety of creative fields.

Where can we find a model or example of the creative process? I think one of the best places to look is Disneyland and other Disney theme parks. More specifically, I believe one of the best models for creativity is found in the design and development of Disney theme parks, a practice better known as Imagineering.

As we'll look at in more detail in the first part of this book, Imagineering was born from the blending of expertise from a number of fields, and just as the first Imagineers adopted techniques and practices from animation and movie-making to develop the craft of Imagineering, we can borrow (and steal) principles, practices, and processes from Imagineering and apply them in other creative endeavors.

I know of few better examples of creativity than Disney theme parks and Imagineering, and I'm not alone in this belief. Garner Holt and Bill Butler of Garner Holt Productions (the world's largest maker of audio-animatronics) write: "Disneyland is still the ultimate expression of the creative arts: it *is* film, it *is* theater, it *is* fine art, it *is* architecture, it *is* history, it *is* music. Disneyland offers to us professionally (and to everyone who seeks it) a primer in bold imagination in nearly every genre imaginable."

I've been a fan and "student of Imagineering" since my first visit to Walt Disney World more than 20 years ago. Even back when I was first learning about Imagineering, I recognized the value it could serve as an inspiration for the creative process, as I wrote in the following review of the first book in my Imagineering Library, *Walt Disney Imagineering: A Behind the Dreams Look at Making the Magic Real*:

> This lavish coffee table book opens the doors on one of the most secret of divisions of the Walt Disney Company, namely Walt Disney Imagineering. These are the people that design, build, and create the various attractions at Disney theme parks and other locations (such as the Disney Store and DisneyQuest). This book explores the process by which the Imagineers conceive, design, and create Disney magic. Anyone interested in the creative process and imagination can benefit from reading this book, if for no other reason than as a source of inspiration for what is possible.

I've been amassing a collection of resources about Imagineering since that first visit to Disney World, trying to learn all I could about how the Imagineers work. In my search to learn as much as I could about this subject, I've identified a set of principles and a process that I believe can serve as a model for the creative process in a variety of fields. I call this set of concepts the Imagineering Toolbox.

The Imagineering Toolbox contains two main types of tools. The first is a set of principles focused on developing and communicating our ideas that I've organized into what I call the Imagineering Pyramid. This was the subject of my first book, *The Imagineering Pyramid: Using Disney Theme Park Design Principles to Develop and Promote Your Creative Ideas*. More specifically, that book looked at fifteen principles the Imagineers use as part of their design process and how those principles can be applied to other creative fields. What the Imagineering Pyramid doesn't address is the process Imagineers use to develop their ideas into real-world parks and attractions. That's the focus of this book, and is the second type of tool in the Imagineering Toolbox. The *Imagineering Process* is a simple process that can serve as a model for developing nearly any type of creative project, from a simple homework assignment to a fully immersive theme park attraction such as Expedition Everest at Disney's Animal Kingdom.

The rest of this book is divided into three primary parts:

Part One: Peeking Over the Berm looks at the origins of Imagineering and the meaning of the word itself, as well as an overview of the Imagineering Process and the evolution of the process. This will give us a foundation on which we can expand in later chapters. We'll also look briefly at the idea of having a "vision" and how that fuels the creative process.

Part Two: The Imagineering Process, the heart of the book, examines the process used by Walt Disney Imagineering in the design and construction of Disney theme parks and attractions. This section contains chapters devoted to each stage in the Imagineering Process. For each stage, we'll look at how it's used by Disney Imagineers, the goals of each stage, and how each can be leveraged in other fields.

Part Three: Imagineering Beyond the Berm explores how to apply the Imagineering Process to a number of specific fields, including game design, instructional design, and leadership and management.

Following Part Three is a Post-Show chapter in which I share some final thoughts, and a few appendices. Appendix A contains a list of the books, DVDs, and other resources in my Imagineering Library (as of this printing, anyway—it doesn't stay the same for long). Appendix B provides an overview of the Imagineering Pyramid, and Appendix C contains a checklist of questions based on the Imagineering Process that you can use when bringing your creative ideas to life.

A Bump Along My Journey into Imagineering

"I am Blue Sky."

The words hung in the air. I felt like I'd been punched in the stomach.

With four simple words, my whole understanding of how the Imagineers designed and built Disney theme parks seemed to be incomplete and incorrect, and I began to question everything I'd learned about Imagineering.

Okay, maybe that's a bit of an exaggeration. In the end it wasn't all that bad, but I did have a few moments of panic.

I suppose some context might be in order.

My wife and kids and I were in the Bamboo Room of the Hollywood Brown Derby at Disney's Hollywood Studios having lunch with an Imagineer. Why were we there? Two reasons.

The first is that I'd been interested in Imagineering since my first visit to Walt Disney World in 1993, and the "Dining with an Imagineer" experience was something I'd wanted to do for a long time. It was a chance talk with one of the people who design and build Disney theme parks.

The second reason is that this lunch was part of my own personal "journey Into Imagineering." As I said, I'd been interested in Imagineering since my first visit to Walt Disney World in 1993, and had started collecting books about the Disney parks and Imagineering in an effort to not only learn more about how the parks are designed, but also to uncover the principles that fuel the Imagineers' creativity. I had come to believe that Imagineering was an ideal model for creativity and the creative process, and that the process and principles the Imagineers use in the design of Disney theme parks could be applied to other creative fields.

This lunch was a chance to dig even deeper than I could through reading books, and gave me the opportunity to ask specific questions about Imagineering from someone whose job it was to design and build Disney theme parks and attractions. I wanted to be able to pick the Imagineer's brain about his processes, techniques, and Imagineering theory. I tried my best to scale back my enthusiasm and not completely dominate the conversation, but I'm not sure how well I succeeded. The lunch exceeded my expectations. Our Imagineer, senior concept designer Jason Grandt, was a gracious host, the food was excellent, and his answers to my questions helped me clarify some of the distinctions I had been making about how Imagineering works. But one of his answers also made me realize that I still had a lot to learn.

How so? I had asked Jason a question and his answer was those four simple words: "I am Blue Sky."

What question could possibly result in that answer, you ask? To answer that, I think I need to give you a little more context.

I mentioned earlier that I had been doing research into Imagineering, specifically how the practices, principles, and processes the Imagineers use in the design of Disney theme parks could be applied to other creative fields. That research focused on two primary areas. The first was around a set of practices and principles that the Imagineers use that I had organized into what I call the Imagineering Pyramid (and the subject of my first book, *The Imagineering Pyramid: Using Disney Theme Park Design Principles to Develop and Promote Your Creative Ideas*). The second area was focused on the process the Imagineers use when designing and constructing Disney parks and attractions. More specifically, I was trying to outline a high-level and simplified version of the Imagineering process.

I had made what I thought was pretty good progress. For example, I knew that Blue Sky and Concept Development were two of the main stages in the Imagineering process, and I also knew that the Concept Development stage happens after the Blue Sky stage. However, I still hadn't made some critical distinctions about the various stages of the process. For example, at that time I thought that the Blue Sky stage was only about brainstorming. As I would soon learn, while brainstorming is an important tool used during the Blue Sky stage, there is more to it than that. I had also identified Design and Construction as the two stages that follow after Concept Development and knew that Models fit into the process somehow as well, but hadn't yet ironed out whether models were a tool used in all stages of the process, or a specific stage in and of itself.

As part of my research, I was also trying to understand in more detail what happens during each stage of the Imagineering process, including where and how specific Imagineers and Imagineering roles fit into that process.

In some cases it seems like a straightforward enough problem, doesn't it? For instance, you might think that show ride engineers and lighting designers would be involved primarily during the Design stage, but it's not quite so cut and dry. They might also be involved from the earliest Blue Sky sessions and continue working on the project right through Construction and opening. The same thing applies to show writers as well as concept artists and show designers. Imagineers of all types are involved at each stage in the process, some in multiple stages, and some even play a slightly different role at each different stage. It was a complex puzzle with lots of moving parts, and I was trying to figure it all out to determine if there were principles at work that might apply outside the parks, or "beyond the berm."

When I walked into our Dining with an Imagineer experience, I *thought* I'd figured it out. But I was wrong.

When I saw that Jason was a concept designer, I thought that meant he worked on Concept Development (they both start with "concept", right?). I asked Jason, "So, as a concept designer, that means that you get involved in the process right after Blue Sky, right?" I was confident I was going to hear something along the lines of "That's right. I work with the ideas generated during Blue Sky and...." But that's not what he said. That was when he hit me with those dreaded words: "I am Blue Sky."

Like I said, a punch in the stomach. And then the self-doubt started. If I had gotten this wrong, what else was I wrong about? Was I kidding myself that I really understood how the Imagineers do what they do? If I couldn't figure out how the process worked, how would I *ever* be able to teach people how to apply Imagineering principles to their own creative ideas?

I don't really remember what happened next. I think his answer surprised me so much that I tried to change the subject. Honestly, I'm not sure how I reacted outwardly—I only know that inside I was feeling deflated and defeated. The rest of the lunch went well enough, and by the time it was over, I had grown determined to figure out where I had gone wrong.

This wasn't the first bump along my journey Into Imagineering nor would it be the last, but it was especially jarring. Fortunately, it was also ultimately an educational "bump" as well. I went back to my Imagineering Library and dug in again, determined to better understand the relationship between Blue Sky and Concept Development and the role (concept) Design played in both. I also wanted to refine my understanding of the entire Imagineering process. It took a lot of re-reading and cross-referencing material in my library, but I eventually began to piece things together and understand Jason's answer. I came to realize that at the time of our lunch, I hadn't yet fully made the distinction between *Concept Design* and *Concept Development*. In my own defense, I think this is actually a fairly subtle distinction, and one I should be forgiven for having made, but that's just me (I talk about this distinction in more detail in Chapter 6: Concept Development).

This stumble wasn't the first time I'd experienced the momentary (and sometimes longer) sensation of not knowing what I was doing. I've experienced many "crises of confidence" in my various careers and jobs and I suspect I'm not alone. I actually think it's a common experience in the creative process. It's something long time Disney film producer Don Hahn talks about in his book *The Alchemy of Animation*:

> The first version of the film is really rough—sometimes embarrassingly rough and difficult to sit through. Then the real process of "plussing" in animation kicks in once again.

It is an iterative process that goes something like this:

- Screen it
- Discuss it
- Get the feeling that you don't know what you're doing
- Weep openly
- Tear it apart
- Correct it
- Re-board it
- Rebuild it
- Screen it again
- Repeat as necessary

It takes five or six times through this process of building and rebuilding before the film is working well enough to go into any sort of production.

I think anyone who has been involved in any sort of creative process can relate to this quote, especially the "Get the sinking feeling that you don't know what you're doing" and "Weep openly" parts. I know I certainly can. This story is just one example of how I've experienced it first hand.

In this case, my questions about the process were my way of screening and discussing my work, and those dreaded four words definitely gave me the sinking feeling that I didn't know what I was doing. Fortunately, I was able to hold myself together without openly weeping (inwardly is a different story), but after I got back home, I tore my notes apart and corrected things, re-boarded, and rebuilt. My next screening was when I presented an early version of the ideas in this book (as well as the ideas in *The Imagineering Pyramid*) at a training conference, and a few years later, in a webinar. After each of those, I repeated these steps again, trying hard to "plus" my work with each iteration.

Which brings us to here and now.

This book is my next screening. I hope you enjoy the show.

PART ONE

Peeking Over the Berm

When you first approach a Disney theme park, it's common that you might see the upper portions of some attractions peeking out over the park's perimeter (known as the "berm"). These exposed structures give guests a peek at what they'll experience once they enter the park. In this section, we're going to take a quick look at Imagineering and the Imagineering Process. Chapter 1 looks at the origins of Imagineering, explores what "Imagineering" really means, and discusses the evolution of the Imagineering Process, starting with Walt Disney. Chapter 2 provides an overview of that process, and Chapter 3 is a discussion of vision and how it informs the creative process.

CHAPTER ONE

What Is Imagineering?

There's really no secret about our approach. We keep moving forward–opening new doors and doing new things—because we're curious. And curiosity keeps leading us down new paths. We're always exploring and experimenting...we call it "Imagineering"—the blending of creative imagination and technical know-how.

—Walt Disney

In this chapter, I want to lay the groundwork for the rest of the book by exploring what the word "Imagineering" means, where the practice of Imagineering came from, and why it's a subject worth studying. This will provide some context for the later chapters, and will hopefully establish a solid footing for our discussion of the Imagineering Process.

What's in a Word?

In technical terms, "imagineering" is a portmanteau (the blending of two or more words into a new word) formed from the words "imagination" and "engineering." Contrary to Disney mythology, Walt Disney didn't create the word Imagineering. The term was used by the Alcoa Corporation in the 1940s in their advertising and marketing to promote what they considered to be imaginative approaches to engineering. But while he may not have coined the term, Walt Disney certainly popularized the word when he used it on his *Disneyland* TV show to promote Disneyland.

Based on its origins, we could think of Imagineering simply as "the combination of imagination and engineering," but that definition falls a little flat for me, particularly when it comes to Disney theme parks. I think it's the word "engineering" that doesn't work for me. While engineering certainly plays a part in Disney theme parks and attractions, the word has technical and scientific connotations not generally associated with other creative or artistic fields that also play significant a part in the design of Disney theme parks.

I think a better definition comes from the quote that opens this chapter. Walt Disney defined Imagineering as *the blending of creative imagination and technical know-how*. I like this definition for two reasons. First, this definition adds the word "creative" to highlight the importance of creativity in imagination. Secondly, the phrase "technical know-how" is a broader term than engineering and encompasses technical and scientific expertise as well as creative and artistic expertise, and is a better reflection of the varied range of disciplines practiced by Disney Imagineers.

The general nature of Walt's definition also means that it applies to a wider range of people and activities. Based on this definition, anyone who's ever used their imagination in the context of some sort of technical know-how has engaged in Imagineering, whether they know it or not. Think about it. The "blending of creative imagination and technical know-how" isn't restricted to theme park design, or even to other so-called creative fields. It can apply to any sort of task, even those that most of us might consider mundane. A file clerk who devises a new filing system can be thought to have "imagineered" the new system. A chef who combines spices and sauces in unique ways might be thought of as a "cooking or culinary imagineer." A science teacher who incorporates games and other activities into the classroom to help students learn is "imagineering" their lesson plan. In short, everyone, including you, has practiced Imagineering in one form or another at some point in their lives.

But the fact that we've all practiced Imagineering isn't what makes it a good model and example of the creative process. There's more to it than that.

To understand why Imagineering is such an effective model for creative projects in other fields, we first have to look more closely at specifically how Disney's Imagineers employ their creative imagination and apply their technical know-how when designing and building theme parks and attractions. In other words, we need to look more specifically at how Disney Imagineers practice Imagineering. We'll explore many aspects of this in later sections, but we need to start with a look at its origins and roots.

Imagineering Origins

Walt Disney Imagineering is the division of the Walt Disney Company responsible for the design and development of Disney theme parks, resorts, cruise ships, shopping areas, and other venues and attractions. Walt Disney Imagineering is the modern name of WED Enterprises (named after Walt's initials, Walter Elias Disney), the organization originally formed by Walt Disney that was created to design and build Disneyland. It was WED that gave birth to Imagineering.

So, how did this organization come to be?

To answer that question, and to understand how Imagineering evolved into the art form and craft that it is today, we need to first go back in time and look at the origins of WED Enterprises and Disneyland.

When Walt Disney first began planning for Disneyland, he approached his friend and noted Los Angeles architect Welton Becket about working on the project. In what has become a famous story in Disney and Imagineering history, Becket told Walt that he should use his own people, because they understood the type of storytelling he was looking to employ at Disneyland. In his foreword to Jeff Kurtti's *Walt Disney's Imagineering Legends and the Genesis of the Disney Theme Park*, Imagineering executive Marty Sklar retells the story this way:

> It [Disneyland] might never have happened if Walt Disney's friend and neighbor, Los Angeles architect Welton Becket, had coveted the design job. For when Walt approached him about designing Disneyland, and explained the concept brewing in his head, Mr. Becket gave his friend this advice: "You'll use architects and engineers, of course, but Walt—you'll really have to train your own people; they are the only ones who will understand how to accomplish your idea.

In 1952, Walt Disney formed WED Enterprises as a company to manage his personal interests and projects, among them the design of Disneyland. Following the advice he got from Becket, the first people Walt brought into WED were artists and art directors from his own studio. Over time he added others, including model makers, architects, storymen, machinists, and others. As the plans for Disneyland expanded, even more disciplines found their way into WED, including sculptors, show writers, special effects engineers, songwriters, landscape architects, lighting designers, and others. When WED took on additional projects, such as designing attractions for the 1964–1965 New York World's Fair and early planning for what would become Walt Disney World, WED grew to include even more diverse types of artisans, craftsmen, and engineers.

Each of these early Imagineers brought with them experience and expertise in other fields that they adapted to the design of Disneyland and other Imagineering projects. Film and animation techniques such as forced perspective, cross-dissolves, and sight gags, once confined to two-dimensional stories told in film and animation, found new expressions in the three-dimensional storytelling that was evolving at Disneyland. Over time, other techniques and practices such as Theming and the use of Wienies were tried, tested, and perfected, adding even more tools to the Imagineering repertoire. Over the years, the various disciplines within WED continued to expand to include electrical engineers, sound design engineers, electronic engineers, ride system engineers, and show control engineers. Today the Imagineers at Walt Disney Imagineering span more than 140 disciplines.

Based on practices and principles borrowed and adopted from these varied disciplines, over time WED developed its own methodologies, techniques, and processes for designing and building attractions. Taken together these have become what we know today as Imagineering. As described by blogger "Merlin Jones" on the *Re-Imagineering* blog on May 18, 2006:

> Out of WED's Imagineering braintrust came the theories, aesthetics, design and engineering of Disneyland, the advancement of three-dimensional storytelling, the development of robotic techniques in Audio-Animatronics and the perpetuation of an "architecture of reassurance" as inspired by Walt Disney's personal sense of optimistic futurism.

In *Walt Disney Imagineering: A Behind-the-Dreams Look at Making the Magic Real*, Kevin Rafferty and Bruce Gordon describe the birth of Imagineering this way:

> Walt and his first team of Imagineers invented the theme park business by inventing the process of "Imagineering." In the course of designing and building Disneyland, the process of "learning and succeeding by dreaming and doing" was employed for the very first time. These new Imagineers used their talents in ways they had never used them before to accomplish things they—or anyone else—had never accomplished before.

The Origins of the Imagineering Process

The Imagineering Process as practiced by Walt Disney Imagineering today is a far cry from the way in which Walt and his original Imagineers did things. Today the process is very formalized, with specific stages and steps, and procedures and approvals for how things are done. In the early days of designing Disneyland, things were done much less formally.

Before we consider the Imagineering Process in detail, let's first look back at early examples of the process and how things were done in the earliest days of WED Enterprises.

In the documentary *Great Big Beautiful Tomorrow: Walt Disney and Technology*, former Imagineer Bob Gurr shares some thoughts on the process in the early days:

> You asked the question, "What was your process like?" I kind of laugh because process is an organized way of doing things. I have to remind you, during the "Walt Period" of designing Disneyland, we didn't have processes. We just did the work. Processes came later. All of these things had never been done before. Walt had gathered up all these people who had never designed a theme park, a Disneyland. So we're in the same boat at one time, and we figure out what to do and how to do it on the fly as we go along with it and not even discuss plans, timing, or anything. We'd just get going on it, and we'd know about when we're going to have to be finished.
>
> It's kind of hard to describe how easy and fast it was working with Walt in those days. In the case of the Autopia car, it was quite obvious that the car needed a little body. I could

see in an instant what was needed, so I made a few sketches, brought them in on a Saturday. They said, "Well, Walt's got a whole bunch more stuff for you to do here," and I said, "Well, I'll just quit my day job and come right over here." Because we were working six days a week and Walt would pay us for five days a week. Try that in today's business world. That's how fast we went, but we were so enthused to design all these new things, and of course it turned out. He wanted antique vehicles on Main Street. I came up with those and pretty soon it was an omnibus, and then a fire engine and after that the Viewliner train, the excursion train. On and on and on with vehicles at Disneyland.

Walt didn't really have what you would say is a formal business plan, or official launch meeting for an official project. He'd walk around and talk to people and say, "Bobby, we're thinking of..." and then he'd suggest something, and then right away we'd see different ways to do that, and then pretty soon there'd be an account number (for the project) that would show up, so we'd officially report our time on something, but there was no project launch. We just worked and Walt just walked around and had suggestions.

Another view on the early days comes from *More Cute Stories, Volume 5* by former Imagineer Rolly Crump. In this collection of stories, Crump describes meeting with retired WED executive Mickey Clark to hear about how Walt built the early dark rides:

Okay, how did this go? And he said, well, "Walt would get an idea," and what we did was, I think we used Peter Pan as an example, and he said, "Walt would get an idea he wanted to do a Peter Pan ride, and he would actually describe the ride. He'd talk about the fact that you would be flying through space, so there'd have to be a conveyance that made you look like you were flying, and then he wanted to make sure the story lines were correct and that you took off from the bedroom and then you went over the city and then finally over the islands, and then finally...you'd finally come back to land."

Anyway, so Walt would ask Mickey, "Now, I want you to get a hold of Dick Irvine and let Dick go to the architect and have the architect decide how big the building was going to be and how much the building was going to cost." He said, "Then I want you to have him go to Roger Broggie and get a cost on the conveyance and the design of the conveyance."

> So basically, he had the whole formula down pat, that he wanted Mickey to find out. So Mickey would take maybe a week or two and do all the homework and everything, and then he would go back and present it to Walt. And if Walt liked what he thought it would take to do it, he would do it.

Both of these recollections highlight that in WED's early days, Walt and his Imagineers didn't follow a specific or formal process. They decided what they wanted to do, and just dove into the work. Over time, as WED took on more and larger projects, such as the 1964–65 New York World's Fair and the New Orleans Square and Tomorrowland expansions to Disneyland, the processes and procedures used by the Imagineers grew more formalized and complex, involving even more people from more disciplines. What's interesting is that even though the process has grown in complexity and formality, at its core it has remained largely the same. In fact, at its core, the process WDI uses today is very similar to the "process" outlined above.

Learning from Imagineering

In *Walt Disney's Imagineering Legends and the Genesis of the Disney Theme Park*, the story of Welton Becket's advice is recounted no less than four times, including in one of the chapter introductions, and with quotes in at least two of the essays about the original Imagineers. Kurtti's recounting of this story emphasizes its importance to the birth of WED and to the craft of Imagineering. For our purposes, the importance is that it shines a light on how we can learn from Imagineering.

Imagineering is a craft based on adapting skills, knowledge, and expertise from one medium and applying them to another. Imagineer Alex Wright describes the parallels between developing animated films and theme park attractions in this excerpt from *The Imagineering Field Guide to Disneyland*:

> The act of designing a "dark ride"... is very similar to that of developing an animated short film. Many of the same techniques are brought to bear, and we go through many of the same steps in the process. We outline our story. We storyboard it. We spend time coming up with gags to make it fun. We integrate the design with any dialog needed to advance

the action. We fine-tune our timing and pacing, and assess the payoff of the story. Even today, the Imagineers rely on the model set out for us by the animators of the Walt Disney Studios, especially those who made the transition from animation to Imagineering and created this new art form.

Imagineering was born from the combined experience and expertise of these early Imagineers, who brought with them technical know-how from film, animation, and other fields and adapted that know-how to the new form of entertainment of which Disneyland was the first example: the theme park. And just as the first Imagineers adopted techniques and principles from film-making when they developed the craft of Imagineering while designing Disneyland, we can adopt the techniques and principles of Imagineering in other fields, such as game design and instructional design.

That's really what this book is all about. It's a close look at the principles behind the process used in Imagineering—the process the Imagineers use in the design and construction of Disney theme parks—with an eye toward applying the same process to other fields "beyond the berm."

CHAPTER TWO

An Overview of the Imagineering Process

The Imagineering Process is a simplified version of the process Walt Disney Imagineering uses when it designs and builds theme park attractions, resorts, and other venues. We'll look at each stage of the process in more detail in Part Two, but for now we'll start with a high-level sketch of each.

There are seven pieces, or stages, in the process. Five stages form the core of the process, while the other two serve as its Prologue and Epilogue.

Prologue: Needs, Requirements, and Constraints

Before the process actually begins, you must identify the project's specific Needs, Requirements, and Constraints. These form the core parameters of your project. The goal of this stage is to *define your overall objective, including what you can do, can't do, and must do when developing and building your project.*

Blue Sky

The initial stage of any Imagineering project is the Blue Sky stage, where initial ideas and concepts are created through a combination of brainstorming and concept design. The goal of this stage is to *create a vision with enough detail to be able to explain, present, and sell it to others.*

Concept Development

Once a concept born from the Blue Sky stage has been selected for development, it undergoes a period of concept development where the idea is more fully fleshed out and realized. The goal of this stage is to *develop and flesh-out your vision with enough additional detail to explain what needs to be designed and built.*

Design

The Design stage is where detailed design documents and specifications are created that will guide the physical construction of the project. These can include architectural documents (blue prints, plans, elevations, etc.) but also more Imagineering-specific documents such as Show Information Guides (documents that outline the story behind the attraction). The goal of this stage is *develop the plans and documents that describe and explain how your vision will be brought to life.*

Construction

The last major stage of the process involves the actual physical construction of the project, including land development and fabrication. The goal of this stage is to *build the actual project, based on the design developed in the previous stages.*

Models

At each main stage in the process, the Imagineers build models of various sizes and scales to help identify and solve potential design challenges. The goal of this stage is to *test and validate the design at each stage to help solve and/or prevent problems that may arise during the design and construction process.*

Epilogue: Openings, Evaluations, and Show Quality Standards

Once construction is complete, the attraction is opened for guests (including cast member previews, soft openings, and finally a grand opening). In addition, once in operation, the attraction is periodically evaluated to ensure that it maintains a level of quality and that the original creative intent of the attraction is intact. The goal of this stage is to *present the project to the audience, allow them to experience it, and evaluate its success and effectiveness over time.*

The diagram below depicts the Imagineering Process and how the various stages relate to one another. As you can see, the Prologue leads to the five core steps of the process—Blue Sky, Concept Development, Design, Construction, and Models—which in turn lead to the Epilogue.

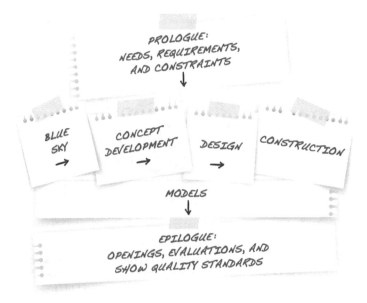

One thing this diagram doesn't show is a feedback loop from the Epilogue back to the Prologue. As attractions and projects are evaluated over time, ideas for enhancements and refurbishments are likely to arise. These become the basis for new Needs (and Requirements and Constraints) which can trigger a new iteration of the Imagineering process. This continuous feedback loop in which the Imagineers look for ways to improve the parks and attractions is an example of what Walt called "plussing" (which is one of the principles in the Imagineering Pyramid).

Is It Really That Simple?

Now, you might be saying, "Is it really that simple? Is it really just those five steps plus a Prologue and Epilogue? There must be more to it than that."

The answer is, "No, it's not that simple, and yes, there is more to it than that".

The Imagineering Process outlined above is a distilled and simplified version of a much more involved workflow. To give you an idea of how involved the actual theme park design process can be, consider the following representative example:

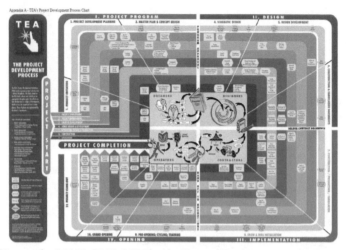

This diagram depicts the TEA (Themed Entertainment
Association) Project Development Process. It has four phases
(Project Program, Design, Implementation, and Opening),
eleven stages (Project Initiation through Project Cost-Out),
seven tracks of activities running through each stage (Project
Development, Project Management, Creative Development,
Venue Design and Development, Attraction Design and
Production, Construction, and Operations), and more activities and milestones than I care to count. In addition, the
process used by WDI adds other steps and stages. As you can
see, it is a complicated process.

So why go from such an elaborate process to something
so simple? There are two main reasons for simplifying this
process as much as I have:

First, the Imagineering Process outlined above focuses on
the primary creative steps in the process. I don't want to get
bogged down with project management and budgeting and
other similar things. These are certainly important and we
sometimes need to deal with them, but for the purposes of this
book, we want to look at the more creative parts of the process.

Secondly, this full process is extremely detailed and complex
(as evident from the above diagram) and is far too much for

anyone to remember, so by reducing it into the five stages plus the Prologue/Epilogue, I'm hoping to make this easier to recall. I'm also trying to employ "Read"-ability, the practice of simplifying complex ideas so that audiences can quickly and easily "read" (understand) them (another principle from the Imagineering Pyramid).

The More Things Change...

In the previous chapter I wrote that even though the process used by the Imagineers grew in complexity and formality over time, at its core it has remained largely the same and that the process WDI uses today is very similar to the process outlined by Mickey Clark to Rolly Crump. As an example, consider the following:

- Walt would come up with an idea for a dark ride: "Let's do a dark ride based on Peter Pan." (Prologue/Blue Sky)
- He would then work out the ride sequence in detail. (Concept Development)
- He would then reach out to various people to determine design details such as the size of the facility, ride system, etc. (Design)
- They then built the ride. (Construction)
- Along the way, Walt had models of everything built. (Models)
- Walt consistently looked for ways to "plus" what he and his Imagineers did. (Epilogue)

CHAPTER THREE

The Creative Process and the Power of Vision

Before we dive in to the individual steps of the Imagineering process, we should first look at a fundamental element of every creative endeavor, and the role it plays in the creative process. Whether you're creating an entire theme park, a single attraction, or something as simple as a short presentation, the core of every creative project is a vision—a picture of what the project will look like when it's complete. A creative vision can be as grandiose as Walt Disney's vision for EPCOT (the Experimental Prototype Community of Tomorrow) or much more modest and simple. In either case, a vision is at the heart of the creative process.

One of my favorite depictions of the power of vision is an image entitled, appropriately enough, "VISION" that depicts a parcel of undeveloped land in Florida (where Walt Disney World would eventually be built), with images of Walt Disney and Cinderella Castle superimposed over the land. The caption for this image is one of Walt Disney's more famous quotes: "It's kind of fun to do the impossible." For me, this one simple (and very "read"-able) image sums up the idea of vision perfectly.

Vision was central to nearly everything Walt Disney created, including his animated masterpieces such as *Snow White and Seven Dwarfs*, his live-action films such as *Mary Poppins*, and what some consider his most significant achievement, Disneyland. Walt had a compelling vision for each project he worked on, and was able to share that vision with the animators, filmmakers, and Imagineers who helped bring his ideas to reality. It was that shared vision that enabled Walt to draw out the best from all the people who worked with him. There have been few creators like Walt Disney who had the ability to develop and share their vision so effectively. I think it's fair to say that Walt's vision and his ability to infectiously share that vision with those around him are the true source of the "Disney magic" that so many fans of his work admire and enjoy.

Walt's vision for what would eventually become Walt Disney World outlived him, and has become a significant part of his ongoing legacy. In *How to Be Like Walt*, Pat Williams and Jim Denney share an often-told story about Walt's vision for Disney World:

> Though Walt envisioned Walt Disney World in Florida, he died before it was built. On opening day in 1971, almost five years after his death, someone commented to Mike Vance, creative director of Walt Disney Studios, "Isn't it too bad Walt Disney didn't live to see this?"
>
> "He did see it," Vance replied simply. "That's why it's here."

A compelling vision fuels the creative process, and can serve as both inspiration and motivation as you work your way through the steps of bringing your vision to life.

Just as Walt Disney's visions for *Snow White*, *Mary Poppins*, Disneyland, and Disney World inspired his animators, film-makers, and Imagineers, a clear vision can help you when you develop your creative projects. If you'll be working with others on your projects, it's important that you share your vision with them and get them excited about it.

Each step of the process that we're going to explore is part of bringing your vision to life, so let's take a quick look at the role of vision in the steps of the Imagineering Process, and what you must do during each step:

Prologue
You define the parameters of your project and identify the problem you're attempting to solve.

Blue Sky
You initially develop your vision and identify what you're going to create. This will include your "story," your creative intent, and other important elements needed to convey it to other stakeholders.

Concept Development
You flesh out your vision, developing the initial ideas created during Blue Sky into more fully formed concepts.

Design
Your vision starts to move toward reality, as you create the detailed design documents (of any number of types) needed to eventually construct and build your vision.

Construction
Your vision becomes real with its actual physical construction.

Models
You create mockups and prototypes at each step along the process, and you start to see your vision take form.

Epilogue
You share what you've created and invite your audience to experience the reality born from your vision. It's also where you step back and evaluate how well you've met your original goals and addressed your original need.

The Imagineering Process

In Part Two, we're going to look at the process the Imagineers use to create Disney theme parks and attractions. The chapters in this section explore each of the stages in the Imagineering process, including how they are employed in the Disney parks, as well as the objective of each stage and how each can be applied "beyond the berm" in other fields and creative projects.

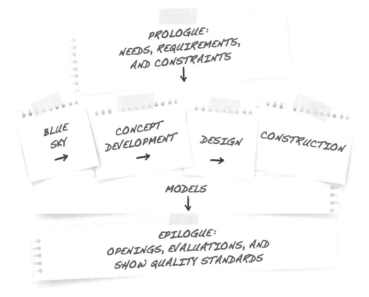

Chapters 4 through 10 each focus on one of the stages of the Imagineering Process: Prologue, Blue Sky, Concept Development, Design, Construction, Models, and Epilogue.

CHAPTER FOUR

Prologue: Needs, Requirements, and Constraints

The initial stage of the Imagineering Process actually takes place before the process begins. The Prologue is very much like a story's prologue that happens before the main events of the story. It's where we identify the specific Needs, Requirements, and Constraints which will serve as the projects' basic parameters.

Needs

Every Imagineering project begins with a Need of some sort, and it is this Need that provides the business and creative motivation for the project. Needs can result from plans for expansion, issues with or evaluations of existing attractions, or even changes resulting from other projects. Projects of all sizes are based on Needs, whether they involve a change to an existing attraction, such as the interactive queues in Haunted Mansion and Big Thunder Mountain Railroad, an entirely new attraction within an existing park such as the addition of Expedition Everest to Animal Kingdom (per Imagineer Joe Rohde, "Animal Kingdom needed a roller coaster, or a thrill ride or some kind"), or even an entirely new theme park such as the addition of Animal Kingdom to Walt Disney World.

A somewhat recent example of a project Need was born out of the expansion to Fantasyland in the Magic Kingdom at Walt Disney World. As part of the New Fantasyland expansion project (which itself was based on a Need to enhance and expand Fantasyland), Mickey's Toontown Fair was closed and the area was changed into an area called Storybook Circus. One consequence of this change is that Mickey Mouse and his friends (including Minnie Mouse, Goofy, Donald and Daisy Duck, and Pluto) were in essence "evicted" from their homes in Toontown. This resulted in a Need for a new permanent home for Mickey and the rest of the group. Disney addressed the need by creating Town Square Theater, located within Exposition Hall just inside the park entrance on Main Street, U.S.A., where guests are able to meet and greet Mickey and his friends.

In some cases, a single project may address multiple Needs. For example, in March 2011, the Walt Disney Company announced a project to address Needs related to the Peter Pan's Flight attraction and to the abandoned Fantasyland Skyway Station (once used by the Skyway ride that carried guests between Tomorrowland and Fantasyland). The queue at Peter Pan's Flight would often become very crowded and occasionally overflow onto the walkways between the attraction and "it's a small world," while the Skyway Station stood largely abandoned for many years, and was used primarily for stroller parking. This project, which took nearly 3 years to complete, moved the restrooms next to the Peter Pan's Flight queue over to the Skyway Station (which was rethemed based on the animated film *Tangled*), and the Peter Pan queue was expanded to include the space formerly used by the restrooms.

Requirements and Constraints

Needs often come with related Requirements and Constraints that impact how the Need can be addressed. Requirements often relate to the creative and operational aspects of how the project will address the Need. These can include things such as creative concerns based on the park or "land" related to the Need or operational concerns such as ride capacity (how many people can experience the attraction in an hour, for instance). Requirements refine Needs in a way that helps the Imagineers create solutions to solve the specific problems which the Needs represent, and help transform them from generic Needs to specific Needs. For example, when designing Expedition Everest for Animal Kingdom, the initial generic Need might have been "Animal Kingdom needs a thrill ride," but additional Requirements refined this into a more specific Need. These Requirements might have been stated as "Animal Kingdom needs a thrill ride that fits with the overall theme of the park," or "Animal Kingdom needs a thrill ride that fits within its Asia land and that can accommodate 2,000 guests per hour."

While Requirements are often related to the story being told by an Imagineering project, Constraints define other types of guidelines, more often those based on "real-world" concerns such as budget, schedule, or other physical limitations that impact how the Need can be addressed. All projects have budgets and schedules that serve to constrain them, and theme park attractions are certainly no exception. Like Requirements, Constraints help refine Needs and define the specific problem to be addressed. The Need for a new home for Mickey and his friends described above provides an example of a time-based Constraint. Given the popularity of Mickey and his friends, the time frame for creating the new home for Mickey and his friends had to be a short one (one to two months at most). With Toontown closing in February 2011, this would mean the new home for Mickey and his friends had to open to guests sometime in March or early April. The Need here might have been stated as something like "We need a new home for Mickey and his friends that can be operational by the end of March 2011."

Another common type of Constraint facing Imagineers are the physical facilities in which new attractions are to be built. This type of Constraint is present whenever an existing attraction is closed and a replacement attraction is planned to be put in its place. Examples of this can be found throughout many of the Disney parks, and include attractions such as Buzz Lightyear's Space Ranger Spin (which is in the facility that once housed If You Had Wings/If You Could Fly, Delta Dreamflight, and Take Flight) and Mickey's PhilharMagic (which is in the facility that once housed Mickey Mouse Revue, Magic Journeys, and Legend of the Lion King). In many cases the Imagineers reuse existing facilities, but in some situations the specific Constraints they face can also lead them to adopt a "scrap-and-build" approach instead. In one specific rumored example, the Imagineers chose to demolish the Horizons pavilion when they built Mission: SPACE because of the Constraints they faced when building Test Track in the facility that had formerly housed the World of Motion.

Another pair of examples will illustrate how Needs, Requirements, and Constraints all play a part in shaping how new attractions are developed.

The current location of Stitch's Great Escape in Tomorrowland in Magic Kingdom at Walt Disney World has been home to four attractions over the years:

- Flight to the Moon (1971 through 1975)
- Mission to Mars (1975 through 1995)
- ExtraTERRORestrial Alien Encounter (1995–2003)
- Stitch's Great Escape (2003 through the present)

In the case of each subsequent attraction, the Need was to update/replace the existing attraction, but each time the new attraction had to fit within the theme of Tomorrowland (a Requirement) and each had to be designed to work within the Constraints imposed by the existing facilities, which includes a theater-in-the-round-style show area.

More recently, the Maelstrom attraction at the Norway pavilion in Epcot was replaced by Frozen Ever After (based on the animated film *Frozen*). Much like the example of Stitch's Great Escape, Requirements and Constraints for this project were based largely on the location of the attraction. Specifically, the new attraction had to fit the theming of the Norway pavilion (a Requirement) and also had to fit in the existing facility (a Constraint). In this case, the new attraction also had to make use of the existing ride layout, including the upwards ramp near the start of the track, the turn-arounds (when the boat switches from moving forward to moving backwards and then forward again), and the drop near the end of the ride. These Constraints imposed significant limitations on what the Imagineers could do with the new attraction.

You Want Me to Do What, Now?

The goal of the Prologue is to *define your overall objective, including what you can do, can't do, and must do when developing and building your project.* One of the quickest ways to fail at a creative project is to not know what you need to create in the first place. Needs, Requirements, and Constraints form the parameters of your project. They define what your project needs to accomplish, how it needs to do so, and any appropriate limitations on the size and scope of the project. These three are critical elements in any creative project, and it's important that you identify all three with enough detail to ensure that you can be successful.

One way to think of these is as follows:

- **NEED**: the problem you're trying to solve ("Animal Kingdom needs a thrill ride"or "I want to write a book about applying Imagineering principles to other creative fields.")
- **REQUIREMENTS**: things you *must* do ("Your research paper must be at least 5 pages in length.")
- **CONSTRAINTS**: things you *can't* do ("You can spend no more than $X on this project.")

As you move into the Blue Sky stage of the process, these three parameters will help keep you on track and make sure that the idea you choose to pursue is appropriate.

Before we dive deeper into Needs, Requirements, and Constraints, let's look at some examples of how they show up in creative projects from beyond the berm:

- A simple example is a school project, such as the standard "five-paragraph essay." Projects of this type have specific Needs (an assigned subject), Requirements (a minimum number of pages/words), and Constraints (a deadline, formatting, etc.).

- A common task in many businesses is putting together a presentation for managers or stakeholders. In this case, the Need is fairly straightforward ("you must put together a presentation for senior management about this issue"). Requirements obviously start with the subject matter, but can also include providing specific recommendations or proposals for addressing the issue, while Constraints can include a deadline for the presentation, as well as a time limit for the presentation itself.

- Tasks in competition and elimination television shows make for interesting examples. One of my favorite shows of this type is *Face Off*, a competition and elimination series focused on special-effects makeup. Each week contestants are given specific instructions for a unique makeup task. For each task, their Need is to create an effective and successful makeup. Requirements often include the theme of the task (horror, science fiction, zombies, etc.) as well as using specific reference and source materials ("your makeup must be inspired by an underwater creature"), while Constraints include a specific time limit (usually 2–3 days) and the makeup and fabrication materials available to contestants in the makeup lab.

IDENTIFYING YOUR REAL NEED, REQUIREMENTS, AND CONSTRAINTS

It's important to make sure you've identified the *true* Need behind your project. What do I mean?

In *The Imagineering Pyramid*, I quoted a well-known marketing essay, *Marketing Myopia* by Theodore Levitt, originally written in 1960 and republished in *Harvard Business Review* in 2004), in which Levitt talks about industries that struggled when they lost sight of their mission and the businesses they were actually in. For example, the railroad industry struggled because railroad companies believed they were in the railroad business instead of the transportation business, while production companies in Hollywood lost market share to television because they believed they were in the movie business instead of the entertainment business.

Something similar to this can happen with Needs. Often times we identify surface Needs that don't fully reflect the real challenge or problem we're trying to solve, and if we don't identify the proper Need, we can spend time and money on a solution to the wrong problem. Authors Clayton M. Christensen, Scott Cook, and Taddy Hall address this idea in the realm of marketing in their January 2006 Harvard Business School "Working Knowledge" article entitled *What Customers Want from Your Products*:

> Marketers have lost the forest for the trees, focusing too much on creating products for narrow demographic segments rather than satisfying needs. Customers want to "hire" a product to do a job, or, as legendary Harvard Business School marketing professor Theodore Levitt put it, "People don't want to buy a quarter-inch drill. They want a quarter-inch hole!"

Taking this idea a step further, in most cases people don't really want a quarter-inch hole, either. What they really want is based on the *reason* they want a quarter-inch hole, such as to hang a picture or shelving. However, if we don't stop and identify the true Need behind the surface, we might never solve the real problem.

When evaluating your project's Need, specifically focus on understanding the "why," as this will help you drill down to the real underlying Need. One approach is to use a variation of the "5 Whys" problem-solving technique, originally developed at Toyota Motor Corporation. "5 Whys" is "an iterative question-asking technique used to explore the cause-and-effect relationships underlying a particular problem. The primary goal of the technique is to determine the root cause of a defect or problem by repeating the question 'Why?' Each question forms the basis of the next question." When used in this context, instead of asking why something happened, you keep asking about the "why" behind your project until you arrive at its true underlying Need.

You should also evaluate and question your Requirements and Constraints. Just like surface Needs can hide or obscure your real underlying Need, sometimes the things we identify as Requirements and Constraints can be based on incorrect assumptions that can lead us in the wrong direction. One source of false Requirements and Constraints is when we continue to do things long after they've served their purpose because those activities become "the way we've always done things." Unfortunately, it's not uncommon for people and companies to lose sight of the true objective of an activity by getting caught up in the details of performing that activity. The challenge is to recognize these activities for what they are and not let them result in false Requirements and Constraints. For instance, if your Need is to develop a new internal process in your business, it might be easy to base your Requirements and Constraints on the ways in which you've traditionally run your business, but unless you examine and evaluate the reasons behind all of those things you've "always done," you could end up creating false Requirements and Constraints that eventually derail your project (see Chapter Six: "Creative Intent" and Chapter Twenty-Two: "Imagineering Leadership and Management" in *The Imagineering Pyramid* for more about the distinction between objectives and activities).

Understanding the "why" behind your Requirements and Constraints is a good way to make sure you've identified real Requirements and Constraints. As with your Need, you can use the "5 Whys" technique to evaluate and find the root cause of your Requirements and Constraints and make sure they are real parameters to consider as you move forward with your project.

EMBRACING CONSTRAINTS

There is a common belief that constraints can be a hindrance to creativity, and that creativity requires total and absolute freedom from limitations and constraints. I don't believe this is true. I believe a certain level of constraint is necessary for creative work. Consider this example. A friend asks you to draw a picture. That's all the instruction you're given. Where do you even start? Should it be a big picture or a small picture?

Should you draw the picture using a pencil, a pen, crayons, or a paint brush? What should the subject of the picture be? The total freedom provided with instructions devoid of constraints leaves you with an almost endless series of questions that can cripple your creativity. But what if that same friend asked you a draw a picture of a house on a sheet of 8.5 x 11 inch paper using a pencil? Wouldn't those constraints make your friend's request a little more manageable? Even simple constraints can make a big difference in the success of a creative project.

In a Bloomberg essay called "Creativity Loves Constraints," former Yahoo CEO Marissa Mayer writes about the value of constraints:

> When people think about creativity, they think about artistic work—unbridled, unguided effort that leads to beautiful effect. But if you look deeper, you'll find that some of the most inspiring art forms, such as haikus, sonatas, and religious paintings, are fraught with constraints. They are beautiful because creativity triumphed over the "rules." Constraints shape and focus problems and provide clear challenges to overcome. Creativity thrives best when constrained.

In *Little Bets: How Breakthrough Ideas Emerge from Small Discoveries*, Peter Sims describes the value of constraints in the context of Frank Gehry's design of the Walt Disney Concert Hall in downtown Los Angeles:

> Instrumental in this process was the way in which Gehry and his team made use of the constraints imposed upon them for the project. On a typical project, the constraints, what Gehry also calls "guard rails," that define the scope of Gehry's figurative box will include a budget, timeframe, materials, political or regulatory rules, and the nature of the building itself. Those constraints not only help Gehry Partners to bound, focus, and measure their progress, they help begin and evolve the design.

Put more simply, Sims notes that "productively creative people use constraints to limit their focus and isolate a set of problems that need to be solved."

When you evaluate the Needs, Requirements, and Constraints of your project, look to how the Constraints you're given can act like Gehry's "guard rails" and help direct you to find a successful idea that meeds your Need.

THE PROLOGUE AND THE PROCESS

Some process models, such as the traditional "engineering design process," often include Needs, Requirements, and Constraints as steps within the process. You may have noticed that the diagram of the Imagineering Process places the Prologue (which contains these three elements) above the five main steps of the process. I opted for this relationship because I view Needs, Requirements, and Constraints as largely external to the main steps of process. Needs, Requirements, and Constraints often come from an external source, such as a customer, client, teacher, manager, etc., and I believe they serve to initiate the process, but are not necessarily part of the process itself. Similarly, the Epilogue is depicted as the result of the process, because like the Prologue, it is external to it.

The Process in Practice: Needs, Requirements, and Constraints

As we move through the stages of the Imagineering Process, I'm going to spend a few words discussing how I went through each stage when writing this book.

The Need behind this book is really more of a want or desire (and for my purposes in this book, wants or desires are effectively the same thing as Needs). After developing the Imagineering Model and my presentation about instructional design, I felt that the principles behind the Imagineering Pyramid and Imagineering Process had a broader application to other creative fields, and I wanted to write a book about it.

Some of my Requirements include publishing details such as the manuscript's format and proper attributions for quotes, but I also have some more "creative" requirements, such as including this "Process in Practice" section and the "Post-Show: Imagineering Checklist" in the chapters that describe the different stages.

Some of my Constraints are based on the medium of the printed word (and in black and white). I need to describe the concepts in the book with words and diagrams (and a few photos). I also have a word count limit and a deadline (both very common Constraints in the publishing world).

Speaking of Constraints, this specific book came about due to a Constraint on my first book, *The Imagineering Pyramid*. I originally planned to include both the Imagineering Pyramid and the Imagineering Process in that book (originally titled *The Imagineering Toolbox*), but for a number of reasons (prime among them a word count constraint), my publisher and I decided instead to pull the Imagineering Process and make it the subject of its own book.

Post-Show: Imagineering Checklist

- The goal of the Prologue is to *define your overall objective, including what you can do, can't do, and must do when developing and building your project.*
- Needs provide the business and creative motivation for the project, and define the problem to be solved.
- Requirements define how a project will address the Need, and what the project *must* do.
- Constraints define what you *can't* do with your project, including budget, schedule, and other physical limitations. Constraints can help narrow your focus and clarify the challenge or problem you're trying to solve.
- Be sure to understand the "why" behind your Need, Requirements, and Constraints. You can't solve problems with creative solutions unless you understand the real problem to be solved.

QUESTIONS

- What is the problem I'm trying to solve? (What is the Need?)
- What are the things I *must* do as part of this project? (What are my Requirements?)
- What are the things I *can't* do as part of this project? (What are my Constraints?)
- Have I thoroughly evaluated my Need, Requirements, and Constraints? Do I know what I *really* need to create?

CHAPTER FIVE

Blue Sky

The first main stage in the Imagineering process is known as "Blue Sky," and it is where ideas for attractions, resorts, and other Imagineering projects are born. In the "Imagineering Field Guide" series, Imagineer Alex Wright defines Blue Sky as:

> The early stages in the idea-generation process where anything is possible. There are not yet any considerations taken into account that might rein in the creative process. At this point, the sky's the limit!

The term "Blue Sky" is actually used in two ways within Walt Disney Imagineering. The first is the name of a specialized department within WDI whose role is to help solve problems that arise during various stages of the Imagineering process. The second, and more pertinent here, is the name of the initial stage in the Imagineers' design process.

In *Walt Disney Imagineering: A Behind-the-Dreams Look at Making More Magic Real,* Imagineer Melody Malmberg describes the Blue Sky stage this way:

> Blue Sky is...the wide-open place where nearly every Imagineering project begins. The earliest phase of a project is always known as Blue Sky, and it can last months or years. This is where big-picture ideas first surface, systems are tested, and assumptions are checked. Here, also, initial budgets are set, before the project moves into the concept and feasibility phase, where even more kinks are worked out in preparation for building.

It is during the Blue Sky stage where Imagineers dream up ideas and initial plans for addressing the Needs that initiated the process. Ultimately, the objective of the Blue Sky stage is to determine what it is that will be built or developed, with enough detail so that everyone understands the proposed scope and intent of the project. The results of the Blue Sky stage don't need to specify every detail of the project (that comes later in the Concept Development and Design stages), but must provide enough detail for high-level budgeting and scheduling estimates to be created.

There are two key aspects to the Blue Sky stage: Brainstorming and Concept Design. Let's look at each of these in more detail.

Brainstorming

Brainstorming is "a group creativity technique by which efforts are made to find a conclusion for a specific problem by gathering a list of ideas spontaneously contributed by group members." Brainstorming sessions are intended to generate as many ideas as possible within a short period of time.

In the context of this book, the primary goal of Blue Sky brainstorming sessions is to generate the sparks that will ignite the fires of the Imagineers' creativity. Richard and Robert Sherman captured this idea in the lyrics to "One Little Spark" from the original Journey Into Imagination attraction at the Imagination pavilion at Epcot:

> One little spark of inspiration
> Is at the heart of all creation

Right at the start of everything that's new
One little spark lights up for you.

All brainstorming sessions begin with a Need, with the goal being to generate as many ideas as possible that address that Need. The focus during brainstorming sessions is centered on addressing the Need rather than on addressing any specific Requirements or Constraints associated with the Need. There will be time during Concept Design and Concept Development to work out how to deal with Requirements and Constraints.

The Imagineers follow a set of rules when brainstorming. In the "Imagineering Field Guide" series, Imagineer Alex Wright outlines these brainstorming rules:

Rule 1
There's no such thing as a bad idea. We never know how one idea (however far-fetched) might lead into another one that is exactly right.

Rule 2
We don't talk yet about why not. There will be plenty of time for realities later, so we don't want them to get in the way of the good ideas now.

Rule 3
Nothing should stifle the flow of ideas. No buts or can'ts or other "stopping" words. We want to hear words such as "and," "or," and "what if?"

Rule 4
There's no such thing as a bad idea. (We take that one very seriously.)

These rules are similar to another set of rules for brainstorming, established by the founder of brainstorming, Alex Osborn. In a *Forbes* article entitled "Brainstorming Is Dead; Long Live Brainstorming," author David Burkus explains that:

> Alex Osborn, the founder of the brainstorming method, set down specific rules for brainstorming, many of which aren't followed in the typical "brainstorming" session. His rules were:
>
> - Generate as many ideas as possible.
> - Defer judgment on all ideas.

- Generate wild ideas.
- Build upon each other's ideas.

There are similarities between the Imagineers' rules and Osborn's rules, but Osborn's include rules about wild ideas and "building on each other's ideas."

Both sets of rules seem focused primarily on ideas and how brainstorming participants should treat ideas generated during a brainstorming session, but what they don't explain is the brainstorming process itself. How does the process start and how do participants know when to stop? How do the Imagineers select specific ideas to pursue and move onto the later stages of the Imagineering process? One answer is found in *Walt Disney Imagineering: A Behind the Dreams Look at Making the Magic Real*, whose authors write that "brainstorming subsides when the basic idea is defined, understood, and agreed upon by all group members."

Another answer to these questions can be found in former Imagineer C. McNair Wilson's *HATCH: Brainstorming Secrets of a Theme Park Designer*, a book that offers a more detailed look at how brainstorming works. *HATCH* explores what Wilson calls his "7 Agreements of Brainstorming." These agreements share principles of both Osborn's and the Imagineers' brainstorming rules, but expand upon these rules to provide a more robust outline of the brainstorming process. Wilson helped promote brainstorming at Imagineering, so the similarities between WDI's rules and his are not surprising.

Let's look at *HATCH*'s 7 Agreements and how they align with both WDI's and Osborn's rules.

Agreement #1: Start a Fire

"Spread the word among the people who will participate in the brainstorming session(s) about the subject matter and the Need to be addressed (similar to how an attraction's Pre-Show introduces the audience to its story and subject matter)." This idea of preparing participants is one that we don't see in either WDI's or Osborn's rules.

Agreement #2: Think Distinctively

"Don't mix or combine creative thinking with critical thinking. Start with creative thinking. There will be time for critical thinking later." This is similar to WDI's rule #2, but clarifies that brainstorming includes both critical and creative thinking.

Agreement #3: "Yes, and..."

"Like the rules of improvisation, use "Yes, and..." to help build upon ideas." This is similar to Osborn's rule #4 about building on each other's ideas, but provides a specific technique to do so.

Agreement #4: No Blocking

"No blocking language." This is essentially the same as WDI's rule #3.

Agreement #5: More Ideas

"Even when you think you can't, come up with even more ideas." This is similar to Osborn's first rule.

Agreement #6: Wild Ideas

"The wilder idea the better. Even the seemingly craziest ideas are worth pursuing, because if they don't end up being used on their own, they can lead to other ideas." This is similar to Osborn's rule #3 about wild ideas.

Agreement #7: Critical Thinking

"After the team has generated ideas via creative thinking and agreements 3, 4, 5, and 6, it's time for critical thinking," which is "about selecting ideas worthy of expanding upon because they contribute to the end result we're imagining together. In brainstorming we talk about critical thinking as focused analytical evaluation, and purposeful thinking with a concise direction, a clear objective to serve the project." Like agreement #1, we don't see this idea in either WDI's or Osborn's rules.

Wilson also gives us a simple approach to applying critical thinking to the ideas generated so far in the process. He calls it "Grab, Group, and Grow":

GRAB

Select an idea to explore. As Wilson says:

> It's not what you like best, most, or if you could only pick one—we're just looking for one idea to start pulling together. Grab one you like...[w]hen you GRAB this idea, you are not deciding this is it, you are merely selecting one theme or idea to brainstorm in a more focused way.

GROUP
Locate other ideas that relate. Wilson tells us to:

> Scan all the ideas on all our lists and locate any and all ideas
> that fit into [your] theme. Start by rediscovering the ideas
> that came as a result of people using "Yes, and ..." when the...
> idea was first suggested. Take all ideas that connect...and
> GROUP them all on the new list.

GROW
Now that you've grouped several ideas, you expand upon them
using the same 7 Agreements of Brainstorming. Says Wilson:

> All Agreements still apply: "Yes, and ...," no blocking/no
> wimping, more ideas (more than just the ones Grabbed and
> Grouped from existing lists), and wild ideas.

After going through the "Grab, Group, and Grow" cycle with
one idea, grab another from your list, and group and grow
again. As Wilson says:

> Repeat the three steps of critical thinking—GRAB, GROUP, and
> GROW—on as many ideas as time and team energy allow. ...
> Even if the first group of ideas or theme is great— overwhelm-
> ingly and enthusiastically embraced by the team—DON'T
> STOP. Grab, Group, and Grow at least three themes.

Eventually the time comes when a decision needs to be
made about which idea to pursue. *HATCH* offers an answer for
this as well:

> If the team continues to return to one theme, one particular
> group of ideas, that one is your most likely answer. That's where
> everyone's imagination, energy, and enthusiasm are repeatedly
> being drawn. Use that inertia to carry over into HATCHing your
> planning and implementation stages to complete the project.
> The team's enthusiasm will also aid you in selling the concept to
> the rest of your company (especially top management).

Of Wilson's 7 Agreements, critical thinking is among
the most important, since it provides a way to take the
ideas created during brainstorming and organize them into
something usable.

THERE'S NO SUCH THING AS A BAD IDEA
The Imagineers' brainstorming rules #1 & #4 emphasize that
there's no such thing as a bad idea during brainstorming,

and the Imagineers truly embrace this precept. Ideas created during brainstorming are recorded and often revisited when new projects make their way through the Imagineering process. As the Imagineers write in *Walt Disney Imagineering*:

> At Imagineering, a good idea is a precious commodity whether developed right away or not. Even when an idea does not make it to the initial design or building stage, it is never forgotten. It may turn up sometime later for use in some other project, in part, or in its entirety....[i]f the spark of an idea is strong, it will never fade away. Even if it travels only far enough to appear on that first piece of paper, there it will patiently remain until the time is right for it to re-ignite.

For example, in an article entitled "8 Key Principles That Disney Imagineers Use to Develop New Attractions," Matt O'Keefe describes ideas originally developed for the Museum of the Weird that were eventually incorporated into the Haunted Mansion:

> Imagineers have great respect for their history, and want to utilize every good idea that their predecessors weren't able to place. A prime example is the Museum of the Weird, a proposed add-on to the Haunted Mansion designed by Imagineer Rolly Crump. After Walt Disney passed, plans to construct the museum were put on an indefinite hold. The Museum of the Weird never came to be as it was originally envisioned, but elements of it were later ingrained into the Haunted Mansion itself, including chairs with faces, a ghostly organist, a seance chamber, busts and portraits that followed you, and paintings that changed right before your eyes. Rolly's design for the enchanted gypsy wagon was even altered to become Madame Leota's Cart.

Other examples include:

- Discovery Bay, a land inspired by the works of Jules Verne and H.G. Wells which included a mockup of the *Hyperion* airship from the 1974 Disney film *The Island at the Top of the World*. Discovery Bay was originally proposed for Disneyland in the 1970s by former Imagineer Tony Baxter. It was never built there, but elements of the plans were later used at Disneyland Paris and Tokyo DisneySea.

- A windmill ferris wheel inspired by the Silly Symphony cartoon *The Old Mill* originally proposed for Fantasyland at Disneyland by Bruce Bushman in the 1950s eventually found a home 40 years later at Disneyland Paris.

Concept Design

The follow up to brainstorming is concept design, the part of the Blue Sky stage where the Imagineers begin to develop ideas generated during brainstorming sessions into concept proposals.

Let's start with an example. Suppose that one of the ideas created during a brainstorming session focused on the Need for a new thrill ride in Animal Kingdom that is something like "a runaway train ride in the Himalayas." Following up on that idea, a concept designer might create a sketch that illustrates what a train ride through the Himalayan mountains might look like. From this initial concept sketch, additional ideas might spring forth ("the train ride should lead guests to a face-to-face encounter with the Yeti!"), leading to additional sketches, models, paintings, and other creations that help the Blue Sky team eventually arrive at a concept for what they eventually hope to build.

The concept design process is often an iterative and generative one, where work on one idea might spawn related (but different) ideas, which in turn can lead to other ideas. The original idea might be ultimately abandoned, but that doesn't diminish its contribution to the project.

It is during concept design that ideas generated during brainstorming sessions are initially filtered based on the Requirements and Constraints related to the Need that was the subject of the brainstorming session. For example, if one of the ideas for the "Animal Kingdom thrill ride" is "a runaway mine car ride through a dragon's lair," that idea would likely have been set aside during concept design (again, ideas are never thrown away at Walt Disney Imagineering) because it doesn't fit within the Asia area or the overall theme of the park (which are Requirements associated with the Need in question).

As an aside: a runaway mine car ride through a dragon's lair might well have been an attraction at Animal Kingdom had the park included the "Beastly Kingdom" land originally proposed to be part of the park, but that's a topic for another time.

Concept designs are expressed in many different mediums and formats, including sketches, paintings, written descriptions, models, and verbal pitches. The goal here is to communicate the concept so that others understand what is being proposed, and the Imagineers use any and all mediums at their disposal to do so. These sketches, illustrations, and other works are intended primarily to convey the overall concept, not necessarily all of the details that will eventually find their way into the project. Imagineer Alex Wright describes these early concept designs in *The Imagineering Field Guide to Disneyland* when explaining the role of various Imagineering disciplines:

> Show/Concept Design and Illustration produces the early drawings and renderings that serve as the inspiration for our project, and provides the initial concepts and visual communications. This artwork gives the entire team a shared vision.

Disney fan Jim Hill shares a conversation he had with Imagineer John Hench about developing initial concept designs:

> I once asked [John Hench] if there was a secret to putting together these early-early concept paintings. And what John said in response kind of surprised me.

> "You can't really be concerned about getting all of the details right. After all—at that point in the project—the design of the park is actually pretty loose, subject to change," Hench explained. "So you have to be a good enough artist to give people the illusion of detail when what you're really doing with an image like this is working in broad strokes."

> Hench added: "You have to understand that the whole point of a concept painting...is that Imagineering is trying to sell some corporation or Disney's board of directors on funding this very expensive proposition. So all your concept painting for this proposed theme park really has to do is convey a sense of energy, the excitement of this place. So accuracy isn't all that important when it comes to creating an image.... What you're really looking to capture here is the sizzle, not the steak."

Hench's comments about "selling some corporation or Disney's board of directors" are of specific interest here. One of the specific goals of the Blue Sky stage is to persuade various stakeholders, whether that be WDI management, park operations executives, or the executive leadership of the Disney company, that the proposed project meets the original Need and is one worth pursuing and investing in.

Specific Blue Sky Outcomes: Story and Creative Intent

Beyond selling an idea to stakeholders, there are two specific outcomes of the Blue Sky stage that are critical to any successful Disney park attraction: the attraction's story (or subject matter) and its creative intent. Story is the fundamental building block of everything Walt Disney Imagineering does when designing and building attractions. The Imagineers identify a core idea for each attraction they build, and it is that core idea, or story, that serves as the basis for every detail of the attraction. The Imagineers also define the attraction's creative intent, or the experience the designer hopes to create for their audience. An attraction's story and creative intent both play a significant role in later stages of the Imagineering Process and in transforming a Blue Sky proposal into a full-fledged Disney theme park attraction.

Beyond story and creative intent, the Imagineers also begin to develop ideas and concepts around other key principles and practices such as:

- Theming and placemaking
- Establishing shots of the attraction and key scenes
- Pre-Shows and Post-Shows
- If and how to use Forced Perspective and Kinetics

These ideas, and others, will be more fully fleshed out and developed in later stages of the Imagineering Process, and we'll consider them in subsequent chapters.

So, I Have an Idea...

In the last chapter we looked at your Need as being the problem you're trying to solve ("we need to create training for our new product") or the thing you want to do ("I want to create a podcast about game design"). Blue Sky is the stage where you develop a vision to address that Need and identify what you're going to create. Put another way, the objective of the Blue Sky stage is to *create a vision with enough detail to be able to explain, present, and sell it to others.*

As you work your way through the Blue Sky stage, remember there are two parts to it: brainstorming and concept design, and both parts play an important role. Through brainstorming you create the spark of your idea, and through concept design you mold, shape, and develop that idea into something you can present to your stakeholders.

While the Imagineers use brainstorming, depending on the nature of your project and your team, traditional brainstorming may not be the best approach for generating ideas. For instance, if you're the only person working on your project (or you're working on a personal project), traditional brainstorming might be difficult (it's not easy to use "Yes, and..." with yourself). In such cases, consider other idea-generation—or ideation—techniques, such as questioning assumptions, using visual/picture prompts, using the SCAMPER (Substitute, Combine, Adapt, Modify, Put to another use, Eliminate, and Reverse) technique, mind mapping, or storyboarding. The goal is to come up with ideas to address your Need, and the goal is more important than the way you achieve the goal. From the set of ideas generated, your last task in this part of the Blue Sky stage is to select the specific idea you wish to pursue.

Once you have selected an idea, you flesh it out and develop it into a concept that you can share with others. This can involve adding details and developing fundamental aspects of the project such as your subject matter (or in Imagineering lingo, your story) and the experience you want your audience to have (your creative intent or objective). Asking yourself questions about these aspects is one way (a good way, in fact) to begin. Use the following questions as a starting point:

- What is your subject matter or story?
- What is your creative intent?
- What is the experience you want your audience to have?
- What is your objective?

Asking yourself questions like these prompts you to provide answers in response and can help you take abstract ideas and make them more concrete. Be sure to write your answers down. Capturing your ideas on paper can help you see them in a different way. Sometimes the simple act of taking the ideas you have swirling around in your head and committing them to paper (or an electronic document or a whiteboard) can help you figure out if the ideas are worth pursuing. It can also help you work out issues or problems that otherwise you might struggle with.

Eventually you need to present your idea to your stakeholders. Depending on the nature of your project, stakeholders can take many different forms, such as the executive leadership of your company or department, your manager or supervisor, your teacher (in the case of a schoolwork assignment), a publisher or client, or maybe just yourself in the case of a personal project. Likewise, your proposal or pitch can take different forms depending on the project. Not all Blue Sky proposals need to be as elaborate as those prepared by the Imagineers. For small and simple projects, your Blue Sky proposal might be as simple as a sentence or two. Remember, the objective of the Blue Sky stage is to create a vision with enough detail to be able to explain, present, and sell it to others.

An important phrase here is "enough detail." Your goal at this stage isn't to work out every detail of every aspect of your project. That comes later in the Concept Development and Design stages. For now the focus should be on the "what"—what you plan to do to address your Need. When presenting your Blue Sky ideas, your focus should be—as John Hench says—to "convey a sense of energy" and "capture...the sizzle, not the steak." A good Blue Sky pitch or proposal outlines how you plan to address the Needs, Requirements, and Constraints of the project with just enough detail to get buy-in from your stakeholders.

Note that the word "sell" here doesn't necessarily mean literally selling your idea to someone else. Like John Hench's comments about "selling some corporation or Disney's board of directors," you need to be able to persuade your stakeholders, whoever they may be, that your idea has merit and that they should continue to fund your work on the project.

During the Blue Sky stage, you may work with several ideas, abandoning some while embracing and developing others. As you work through ideas in this stage, evaluate the concept and vision of each against your original Need, Requirements, and Constraints. This can help you validate your ideas before you invest too much time and effort moving in the wrong direction.

IN DEFENSE OF BRAINSTORMING

In recent years brainstorming has gotten a bad rap in some business circles. From concerns about extroverts dominating brainstorming sessions, to studies suggesting that individuals working independently can generate more ideas than teams working together, criticisms of brainstorming have become more common. In his *Forbes* article, Burkus outlines some of the criticism lodged against brainstorming, but points out:

> What our experiences and the critics of brainstorming fail to realize is two-fold: 1) brainstorming is typically never done properly, and 2) brainstorming isn't a stand-alone process.

The Imagineering Process addresses both of these concerns. Regarding the first, investing the time and effort to adopt WDI's and Osborn's brainstorming rules, as well as *HATCH's* 7 Agreements of Brainstorming, can ensure that your Blue Sky Brainstorming is done "correctly," and is productive and fruitful. Speaking about his second point, Burkus emphasizes the importance of other stages in the creative process:

> In addition, brainstorming as an idea generation method isn't a stand-alone process. Brainstorming represents an exercise in divergent thinking, but combining ideas and applying a convergent thinking process is just as important. This is why every major creative process involves some idea generation stage like brainstorming, but also stages that evaluate, prototype, and implement ideas. Very rarely does a market-changing product or a groundbreaking innovation look like any of the ideas that came up during a brainstorming session. The end result of brainstorming is a list of ideas that may or may not solve the problem at hand. If you're looking just at that list, and no idea jumps out as the perfect solution, it's easy to believe the session was a failure. That's what makes other stages so necessary.

As we'll see in later chapters, the Imagineering Process includes stages that involve evaluation (Concept Development and Design), implementing (Construction), and prototyping (Models).

The Process in Practice: Blue Sky

For this book, I started with a simple Need: "I want to write a book about applying the Imagineering Process to other creative fields." Starting from this, for the Blue Sky stage I needed to flesh this out just enough so that my stakeholders (myself and my publisher) understood what the book would be when it was published. I didn't really do much brainstorming. Most of my work in this stage was working out more specifics about what I intended to write about the process both inside the parks and beyond the berm. Another way of saying this might be:

> *The Imagineering Process* explores the process used by Walt Disney Imagineering in the design and construction of Disney theme parks, and how the principles at work in this process can be modeled and applied to creative fields that lie "beyond the berm."

Post-Show: Imagineering Checklist

- The objective of the Blue Sky stage is to *create a vision with enough detail to be able to explain, present, and sell it to others*.
- A good Blue Sky proposal outlines how you plan to address the Needs, Requirements, and Constraints of the project with just enough detail to get buy in from your stakeholders.
- Brainstorming *isn't* just about generating lots of ideas. It's also about identifying potential ideas, and grouping together similar ideas, and growing them.

QUESTIONS

- Are you using Wilson's 7 Agreements of Brainstorming? WDI's rules? Osborn's rules?
- What other ideation techniques could you use if brainstorming isn't appropriate?
- Do you have a vision for your project?
- Have you developed your vision enough to be able to explain it to others?
- Have you developed your idea enough to convince your stakeholders (whoever they may be) to proceed with your project?

CHAPTER SIX

Concept Development

The next stage in the Imagineering process is Concept Development. As its name implies, Concept Development, also referred to as simply "Concept" by WDI, is where the initial concept created in the Blue Sky stage is further developed and fleshed out enough so that more detailed design work can begin.

The Imagineers describe this move from Blue Sky to Concept Development and the role of this stage in *Walt Disney Imagineering*:

> Brainstorms provide our spark with a solid foundation upon which we build a definitive design. Though we can imagine the end result, the concept is a long way from reality. Our next step involves a little more imagination—and a lot more blank paper—as we define the details of the idea, and determine how it can best emerge to tell its story in a three-dimensional world.

As this quote suggests, developing details about a given concept or idea is the core of Concept Development. More specifically, during Concept Development the Imagineers enhance, extend, and embellish every aspect of their concept until they've defined it well enough that real design work can begin.

Expedition Everest at Disney's Animal Kingdom provides a great example of how Blue Sky concepts are fleshed out and embellished during Concept Development. Let's suppose that the concept for Expedition Everest created during Blue Sky was something like "A runaway train ride through the Himalayas that culminates in an encounter with the Yeti." That concept, along with concept drawings and models, may be enough to explain what the attraction will be at a high level, but it's still fairly vague and doesn't provide nearly enough detail to move into the Design stage. Consider the following questions left unanswered by that initial concept:

- Who are the guests in the story?
- What specifically happens along the runaway train ride?
- What are some of the specific scenes that comprise the attraction?
- What is the background story of the train ride?
- What is the experience we want the audience to have on this attraction?
- Where does the attraction's story take place?
- Why are guests aboard the train in the first place?
- When does the attraction take place (in what time period is it set)?
- What does the Yeti look like?
- What role does the legend of the Yeti play in the attraction?

Answers to these (and other) questions lead the Imagineers to flesh out and develop the concept with the additional detail they need. In some cases, the answers to some questions lead to other questions, and so on. (I admit that the initial concept I'm using for this example is simple—and most Blue Sky concepts are probably somewhat more developed than this—but the principle at work still applies.)

To continue the example, answers to the above questions provide additional details about the attraction, such as the name and back story of Serka Zong, the village that surrounds it; the design of the queue and the idea that it takes guests around the Yeti Mandir and through a museum devoted to the Yeti; the name and background of Himalayan Escapes—Tours and Expeditions, the expedition firm and railroad company that will lead guests up toward the Forbidden Mountain and Mount Everest (and ultimately to their meeting with the Yeti); and the story of the doomed expedition that guests learn about as they make their way through the Yeti museum. These details give the concept more depth, and give the Imagineers more substance to work with as they develop the attraction.

You may notice that nearly all of these details are focused on the attraction's story. As noted in the previous chapter, the Imagineers identify a core idea for each attraction, and it is that core idea, or story, that serves as the basis for every detail they develop during this stage. Story is a primary area of focus during Concept Development, and is the fundamental building block of everything Walt Disney Imagineering does.

The focus on story isn't limited to details of the story that guests will find themselves immersed in. It can also include fleshing out the background and history of the attraction. An attraction's back story can provide details the Imagineers use to further develop the attraction. For example, consider the following background story elements for these attractions:

Big Thunder Mountain Railroad
- Details about the cursed mine
- History and details about the town of Rainbow Ridge
- Details surrounding the tragedy that struck the area

Mission: SPACE
- History of the International Space Training Center (ISTC)
- Details about the planned mission to Mars
- Details about the X2 Deep Space Shuttle

Twilight Zone Tower of Terror
- History of the Hollywood Tower Hotel
- Details of the night of the tragedy

Beyond story, other specific areas of focus for the Imagineers during this stage include creative intent and characters. An attraction's creative intent is the experience the designer hopes to create for their audience, and what the designer wants to accomplish with the project. A big part of bringing an attraction's creative intent to life involves mapping out the experience the audience will have on the attraction, including specific scenes or encounters that guests will face along the way. The Imagineers use storyboards to help work out the sequence of events guests will experience, starting with the attraction's queue and pre-show all the way through its post-show and exit. Going back to our example of Expedition Everest, some specific scenes guests will experience include:

- Traveling up the mountain and through the Yeti shrine
- Encountering the broken train tracks near the peak of the Forbidden Mountain
- Racing backwards through the mountain in pitch darkness
- Witnessing the Yeti tearing up a section of track
- Barreling along the tracks outside and around the mountain
- Encountering the Yeti face to face as it lunges out at the passengers

Along with story and creative intent, the Imagineers also focus on creating and developing characters during this stage. As the Imagineers explain, "Once a ride or show concept has been determined, the next step we take is to 'populate' the newly created world." The characters that populate theme park attraction must be designed to ensure that audiences can understand them quickly and easily. "Because of the relatively short length of a theme park attraction, all of the ingredients that make up a character and his or her role in the show must be recognized almost immediately."

The Imagineers also focus on more practical aspects of their projects during Concept Development, such as the attraction's queue, its pre-show and post-show, and if applicable, any related merchandise locations (which might be an "exit through retail" post-show). Beyond these, the Imagineers also have to consider how the attraction will fit within its eventual home and with other attractions around it. For instance, when working on Expedition Everest, the Imagineers had to work out how it would fit into the rest of the Asia area of Animal Kingdom, as well as how guests would approach the attraction as they make their way through the park, both when approaching from Africa and from DinoLand, U.S.A. The Imagineers also flesh out their ideas and concepts around other key principles and practices such as Theming and Placemaking; Long, Medium, and Close Shots; Wienies; Transitions; Forced Perspective; Kinetics; and "Read"-ability. (These concepts are described in detail in *The Imagineering Pyramid*.)

Concept Development is the stage where the Imagineers do whatever research is needed to bring the attraction to life. While some research takes place during Blue Sky, the majority of it occurs during this stage. In *The Disney Mountains: Imagineering at its Peak*, Jason Surrell describes the role of research in the development of Expedition Everest:

> Disney attractions have been marked by exhaustive research and meticulous attention to detail from the very beginning, but Expedition Everest took that philosophy to new heights of immersion, pun very much intended.

In the case of Expedition Everest, the Imagineers made multiple trips to Tibet, China, and Nepal while researching the Himalayas and the legend of the Yeti. This helped in their designs of Serka Zong, the village nestled at the bottom of the Forbidden Mountain, as well as the abandoned tea plantation that now serves as the base of operations for Himalayan Escapes—Tours and Expeditions.

Concept Development is also the stage where project planning begins to play a part. While initial budgeting and estimating might be done as part of the Blue Sky stage, they occur in earnest during this stage, since it is here where the project begins to become real in terms of timelines and detailed project plans. The goals of this type of planning are to not only map out the size and scope of the project, but also to determine the feasibility of the proposed project (see "Feasibility: Between Concept Development and Design," below). Project planning during Concept Development is also where the

Imagineers factor in other Requirements and Constraints that may have been glossed over during the Blue Sky stage.

The output of the Concept Development stage is similar to the output from Blue Sky, and can include paintings, drawings, models, text pieces, storyboards, and other items. The Imagineers produce more of these during Concept Development than during Blue Sky, since this stage involves drilling deeper into the project. For example, a handful of "big picture" concept renderings might come out of the Blue Sky stage, but concept art for every portion of the attraction is developed during this stage. Perhaps the most widely known—and among the most important—of these outputs are concept paintings, described by the Imagineers as:

> A concept painting is an open window to the future, offering an exciting glimpse of the concept as it might some day appear. These works are completed during the earliest stages of a project's development, loosely based on a smattering of fresh and unrefined information. Yet, they are often used later as a source for color and material selection, and as inspiration for model building, set, landscape, and lighting design, and even architectural plans and elevations.

The Imagineers create a wide variety of concept paintings and illustrations during Concept Development, including overall concepts that illustrate how an attraction or venue will fit into the area around it, paintings that show specific views of the outside of an attraction (exteriors) or illustrate the atmosphere inside an attraction (interiors), and paintings that provide details of specific scenes within the attraction. As the quote above suggests, detailed concept paintings like these play a major role in later stages of the Imagineering Process, providing inspiration and guidance for more detailed types of design documents. Blueprints and other technical design documents can't be developed without finished concept artwork. In *Dream It! Do It!*, Imagineering executive Marty Sklar tells a story that illustrates this:

> One day, DIck [Irvine] gave me a new assignment: get Herb Ryman to finish the concept design for the Walt Disney World castle. ... "Herb is holding up the whole project," Dick explained. "The architects can't do the design and working drawings until they have a concept direction."

Concept Design vs. Concept Development

You might be asking, "What is the difference between Concept Design and Concept Development?" Well, if you are, you're not alone, because it's a fairly subtle difference. When I first starting working on ideas for this book, I didn't understand the difference myself, and it was only after my family's lunch with an Imagineer that I started to understand the distinction between the two (I describe this story in more detail in "Pre-Show: A Bump Along My Journey into Imagineering," above). Concept Design is concept design/development work done during Blue Sky where the ideas from brainstorming sessions are fleshed out and developed into project proposals. Concept Development is about taking those ideas and further developing them such that real design work and project planning can be done to turn the idea into reality.

FEASIBILITY: BETWEEN CONCEPT DEVELOPMENT AND DESIGN

In this high-level look at the Imagineering process, I've omitted some stages along the way to keep things simple. One such stage is Feasibility, where the feasibility of the proposed project is evaluated, based on a number of factors including cost (it would cost How Much?), technology (we would need to build What?), and thematic fit (and just how does that idea fit into this park?). Many project proposals never make it past Feasibility, and are sent back into Blue Sky and Concept Development for additional development and rework. If "the sky is the limit" during Blue Sky, feasibility might be thought of as the force of gravity pulling the project back down to earth.

Taking Your Ideas to the Next Stage

In the Blue Sky stage you figured out what you want to do and created a vision for your project, but before you can begin to actually execute and implement your idea, you first need to develop that concept until you really understand what it is you plan to create. This is what Concept Development is all about. The goal of the Concept Development stage is to *develop and flesh out your vision with enough additional detail to explain what needs to be designed and built.*

Designing theme park attractions often involves extensive Concept Development work and results in dozens and dozens of paintings, drawings, and other documents, but creative projects beyond the berm can vary considerably in size and scope. For instance, if you're working on a school assignment, such as a presentation, Concept Development might involve little more than creating a detailed outline of the topics you intend to cover in your presentation. If you're designing a series of e-learning training courses for an enterprise software application, your Concept Development work may entail several extensive outlines as well as prototypes or proofs of concept for the interactive elements of your training. Again, the goal is to develop your idea with enough additional detail so you know what you need to design and build.

So, how do you go about developing and fleshing out your concept or vision? There are several possible approaches, but among the most effective tools is asking and answering questions. Blue Sky concepts are often "big picture" concepts and are somewhat vague and short on detail, and as such leave room for lots of unanswered questions, such as "What about *this*?", "What about *that*?", "Who is the audience?", etc. Questioning various aspects of your concept is a good way to learn what you don't yet know about it. Each question you can ask points you to some new idea or insight that can help you develop your concept. In some cases, the answers to some questions lead to even more questions, and so on.

A good place to start when coming up with questions to ask is to look for aspects of your concept that are vague or undefined. For example, if you're designing training for an enterprise software application, during Blue Sky you might summarize your concept as "a set of training courses that address installation, implementation, and configuration, and ongoing use of the application." That description is pretty vague, and doesn't provide enough detail for you to start designing the training; you need to know a little more about the courses you plan to build first. Some example questions that can help you develop this concept might include:

- What specific topics should each course cover?
- Who is the audience for each course?
- What are the steps involved in installing this software?
- What does someone installing the software need to know beforehand?
- What are the specific functional areas of the software that require configuration before they can be used?
- How do users navigate and work with the software?
- Are there different types of users, and does each type require different types of training?

In addition to specific questions about your concept, you can also ask yourself more general questions during this stage to help make sure your Concept Development work is taking you in the right direction. The principles outlined in *The Imagineering Pyramid* are one source of questions of this type, such as:

- How can I strengthen and emphasize the subject matter or story? (It All Begins with a Story)
- How can I refine and enhance the audience's experience? (Creative Intent)
- What sorts of details can I use to reinforce my subject matter? (Attention to Detail and Theming)
- What are our establishing, medium, and close shots? (Long, Medium, and Close Shots)
- How can I attract the audience's interest and capture their attention? (Wienies)
- What is the best sequence in which to communicate the subject matter? (Transitions)
- How can I introduce and reinforce our subject matter? (Pre-Shows and Post-Shows)
- Where can I use the illusion of size to help communicate my message? (Forced Perspective)
- Are there ideas and concepts that I need to simplify? ("Read"-ability)
- How can I keep the experience active and dynamic? (Kinetics)

- Where can I employ repetition and reinforcement? (The "it's a small world" Effect)
- How can I engage our audience? (Hidden Mickeys)
- How can I make this better? (Plussing)

One specific question you should ask during Concept Development is: "Does my concept have any assumptions that I should re-evaluate?" As we talked about in Chapter 4, it's useful to evaluate and question your Needs, Requirements, and Constraints so that you really understand the "why" behind them. The same holds true for other aspects of your concept. Part of developing your ideas is questioning the decisions you've made thus far to make sure they are still valid. Developing your ideas based on incorrect or unnecessary assumptions can lead you in the wrong direction.

A great example of this comes from the story of the development of the Apollo Lunar Module (or LM), told in the "Spider" episode of the HBO mini-series *From the Earth to the Moon* and in the book *Moon Lander: How We Developed the Apollo Lunar Module* by Thomas J. Kelly. The initial proposal design for the LM (i.e., the Blue Sky design) included seats for the astronauts as well as large windows that would allow them to see where they were flying when landing on the Moon. Weight was a major constraint for the LM (it had to be transported to the Moon after all) and one of the designers' main goals during the Preliminary Design stage (the equivalent of the Concept Development stage for the project) was to reduce the overall weight of the vehicle. The large cockpit windows each weighed several hundred pounds, but couldn't be removed because the astronauts needed to be able to see from their seats. It wasn't until an engineer asked the question "What if they don't need seats?" that the LM designers realized there was another solution. By removing the seats and having the astronauts stand, the astronauts were closer to the windows and thus had a larger field of vision and could make do with much smaller windows. The incorrect assumption that the astronauts *needed* seats had led the designers in the wrong direction (it's doubtful that the LM would have made it to the Moon based on its original design).

Research is another useful tool during this stage. Just as the Imagineers do field research during their Concept Development stage when creating new attractions, this is the stage when you must do whatever research you need to fully understand how you can design and build your project. Research can take many forms, including identifying online resources such as articles and websites, reading books, listening to podcasts, watching videos, and even conducting interviews. If you're not an expert in your subject matter, an effective part of your research could be to identify and interview subject matter experts (or SMEs) who can help you better understand your subject matter. In my experience, more often than not one form of research leads to another, such as when you learn about a new book while listening to a podcast interview or reading an online article. Good research takes many forms, and uses many different sources. Imagineer Joe Rohde echoes this sentiment with his Instagram followers in a post about researching Mayan headdresses dated on May 28, 2017, in which he writes "To designers. Research. ... [a] single image is never enough for good research. You need many many images, as many as you can get."

And while the bulk of your research should happen in this stage, there is a role for research in every stage of the Imagineering Process. In Chapter 4 we talked about identifying your "real" Need, Requirements, and Constraints, which often requires some level of research. Likewise, your Concept Design work in the Blue Sky stage may require you to do some research to make sure your concept is a workable one. As author Vibeke Norgaard Martin notes in *101 Things I Learned in Law School*, "Research is a primary, not preliminary, activity... [r]esearch isn't finished until the deadline arrives."

You should also use the Concept Development stage to revisit any Requirements and Constraints that you haven't fully addressed yet. Not all Blue Sky designs will address every Requirement and Constraint, so it's important to that you return to them during this stage to make sure you have answers to the challenges they present. As was the case with Blue Sky, part of the goal of Concept Development is to persuade your stakeholders to continue to support (and where appropriate) fund your work so you can move on to the next stage. One of the surest ways to be stopped in your tracks when working on creative projects is to not address Requirements and Constraints. To return to the Apollo Lunar Module example, the Grumman Aircraft designers couldn't move on to their Detailed Design phase had they not found ways to reduce the weight of the spacecraft to within the project's constraints.

In terms of output, the results of your Concept Development work can include outlines, storyboards, drawings and diagrams, models, mockups, and prototypes. You should plan to create whatever you need to fully communicate your fleshed-out and developed concept. How much detail is enough? A good rule of thumb is to ask yourself, "Do I have enough information to actually build this project?" If not, keep asking more questions and doing more research until you do. Another good question here is, "Have I developed my concept enough that I could turn it over to someone else and they could take the next steps?" If the answer is no, you're not quite ready to move on to the next stage.

The Process in Practice

For this book, my Concept Development work primarily focused on fleshing out my concept for the book and creating an outline that provided a structure for organizing the information I wanted to cover. A book about "applying the Imagineering Process to other creative fields" is a fairly vague description of what I wanted to write. I needed to break that concept down and consider how I might go about explaining how to apply the Imagineers' process to other fields. Some of the questions I needed to ask and answer in this stage included:

- What is my Creative Intent?
- What do I want readers to come away with after reading the book?
- What are the stages of the Imagineering Process?
- How do I want to describe and explore each stage?
- How I can present the Imagineering Process in terms of other creative fields?

I used mind maps for outlining, which allowed me to step back and see the entire book and work out the best order and sequence for my various sections and chapters. I even ended up creating mind maps for some of the individual chapters which I used to experiment with the order of topics I wanted to cover until I settled on the final version you're reading (the Blue Sky chapter was one I spent considerable time on).

Post-Show: Imagineering Checklist

- The goal of the Concept Development stage is to *develop and flesh out your vision with enough additional detail to explain what needs to be designed and built.*
- The scope of the project determines the extent of your Concept Development work.
- Questions are an effective tool for finding areas and aspects of your concept that need further development.
- Concept Development is the stage where you learn what you don't yet know about your project.

QUESTIONS

- Have you developed your concept enough that you could turn it over to someone else and they could take the next steps?
- Are you using questions to find aspects of your concept to further develop?
- Is your concept feasible?
- Have you done enough research?
- Does your concept have implicit assumptions that could be questioned/challenged?

CHAPTER SEVEN

Design

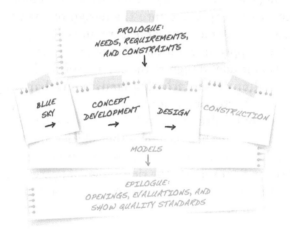

PROLOGUE:
NEEDS, REQUIREMENTS,
AND CONSTRAINTS

BLUE SKY → CONCEPT DEVELOPMENT → DESIGN → CONSTRUCTION

MODELS

EPILOGUE:
OPENINGS, EVALUATIONS, AND
SHOW QUALITY STANDARDS

When a project moves past the Concept Development stage (as well as the Feasibility stage, see "Feasibility: Between Concept Development and Design"), it moves on to the Design stage. Also referred to as the "Schematic" stage by the Imagineers, this stage is where the vision for the attraction depicted in concept art is translated into detailed plans that can be used in its construction.

At a high level, there are three main areas of focus during this stage: Facility Design, Ride Design, and Show Design. As their names suggest, these address the design of the physical facility that will house the attraction, the ride system used in the attraction (if applicable), and the attraction's story and thematic elements. Let's look at each of these in a little more detail.

Facility Design involves the design of the physical building(s) that will house the attraction. This includes not only the building itself, but also the infrastructure to support the building and any utility systems needed by the attraction, such as plumbing, electrical systems, gas/fuel lines, HVAC, etc. Some of these are fairly obvious, such as HVAC systems (any building that guests will enter is likely to have air conditioning) or the plumbing requirements of water rides such as "it's a small world," Pirates of the Caribbean, and Splash Mountain, but other attractions might have less obvious utility requirements. For example, the theater effects of Mickey's PhilharMagic at Magic Kingdon require that plumbing and electricity be incorporated into the design of the theater's seating, and any attraction that features fire will require some sort of fuel supply. In addition to interior concerns, Facility Design also includes design of the exterior of the facility and its surrounding areas, including landscaping/hardscaping, walkways, and exterior queues.

Ride Design involves design of the ride system used by the attraction, if applicable, including the vehicles and the track on which the vehicles will travel, as well as the systems used to control them. For example, the ride design of Space Mountain included design of the computer systems that track the relative locations of all of the vehicles on the ride track. Ride design often involves adopting new technologies or specific applications of existing technologies. For example, the ride systems of the Tomorrowland Transit Authority PeopleMover at Magic Kingdom (originally known as the WEDWay PeopleMover) and the Rock 'n' Roller Coaster Starring Aerosmith attraction at Disney's Hollywood Studios use linear synchronous motors, while the Twilight Zone Tower of Terror involves use of an elevator system that actually pulls the car downwards faster than the force of gravity.

Show Design involves design of the sets, props, decorations, and theming that bring the attraction's story to life. Show Design is the part of the Design stage where iconic Disney park scenes such as the Wench Auction and the Jail scene in Pirates of the Caribbean, the Grand Hall and Seance Room in the Haunted Mansion, and the burning of Rome in Spaceship Earth are brought to life. In addition to the sets and scenes in the attraction itself, Show Design also includes design of the attraction's queue as well as its pre-show and post-show (if applicable).

Show Design is also the stage where the Imagineers put many of the Imagineering techniques and principles described in *The Imagineering Pyramid* into practice, including:

- Using Attention to Detail and Theming to reinforce the attraction's story and creative intent
- Using Long, Medium, and Close Shots to draw guests into the story
- Creating Wienies to attract guests' attention
- Designing appropriate Transitions between scenes and areas
- Using Forced Perspective to make objects appear larger or smaller as needed
- Designing show scenes with "Read"-ability
- Employing Kinetics to help make scenes more active and dynamic
- Reinforcing ideas and concepts through repetition and other means (The "it's a small world" Effect)
- Incorporating Hidden Mickeys
- Identifying areas for Plussing

While each of these three areas (Facility Design, Ride Design, and Show Design) stand on their own in some respects (for instance, the specifics of the HVAC system will likely have little to do with the design of a ride vehicle), each of them influences—and is influenced by—the other other two. For example:

- Ride and Show considerations need to be taken into account when designing the facility, which needs to accommodate the ride system, as well as all of the space needed to build the show sets and scenes.
- Facility and Show considerations need to be taken into account when designing the ride system, which has to function within the facility's design as well as intersect with and pass through various show scenes and sets.
- Facility and Ride system considerations need to be taken into account when designing the Show elements. Show scenes and sets need to fit within the facility and accommodate the ride system as appropriate.

There are other more specific ways in which Facility Design, Ride Design, and Show Design intersect with each other. For example, designing ride vehicles for attractions such as Expedition Everest, Seven Dwarfs Mine Train, or the Slinky Dog Dash in Toy Story Land at Disney's Hollywood Studios (planned to open in 2018) involves a combination of Ride Design and Show Design. The Ride Design focus is on the design of the mechanics of the physical vehicle, while the Show Design focus is on the decorative and thematic aspects of the vehicle.

Similarly, designing buildings and facilities involves both Facility Design and Show Design, where the design of the physical buildings has to support the show elements. The new Pandora area at Disney's Animal Kingdom provides a good example; its Facility Design had to accommodate and incorporate major show elements such as the floating mountains and the local flora and fauna found in the Valley of Mo'ara.

Another good example of the intersection of Facility Design and Show Design is (or was) Mickey and Minnie's houses at the former Mickey's Toontown Fair at Magic Kingdom (which closed in 2011 and was replaced by Storybook Circus), where the cartoon nature of the buildings resulted in the Show Design dictating the design of the physical buildings. In *The Imagineering Field Guide to the Magic Kingdom at Walt Disney World*, Imagineer Alex Wright touches on this in his description of the "squash and stretch" animation technique and how it was used in Mickey's Toontown Fair:

> As it relates to architecture, squash and stretch is expressed in several ways. You'll notice that there are very few straight lines in Mickey's Toontown Fair. Mickey's and Minnie's houses appear to be, though they're not, sort of soft to the touch, like pillows. The props and set-pieces—such as the cars, planes, tools, and televisions—all take on a cartoon-y character, almost as though they were inflated from within.
>
> The squash-and-stretch approach actually led to some difficulties in getting the buildings built. Construction workers have a tendency to want to make their buildings straight and true, the way people usually want them to be, so we had to work hard with them to get across the notion of how to build a *squashed* house. All those curves were necessary, however, to make Mickey's Toontown Fair a place where the characters

could really live. This way, the architecture relates to the characters, and the characters relate to the architecture.

The work in each of these three areas is continuously shared and reviewed by teams in all three areas during this stage to ensure that the designs are cohesive and fit together to bring the attraction to life.

Imagineering Disciplines

If we think of Facility Design, Ride Design, and Show Design as forming the macro level of the Design stage, the micro level comprises dozens and dozens of various and varied Imagineering disciplines working together. As Alex Wright explains:

> Imagineers form a diverse organization with over 140 different job titles working toward the common goal of telling great stories and creating great places. WDI has a broad collection of disciplines considering its size, due to the highly specialized nature of our work.

During the Design stage the Imagineers employ a multitude of design efforts that involve Imagineers from several disciplines. It's during this stage that many of WDI's 140 disciplines come into play. Let's look at some of these disciplines:

SHOW SET DESIGN breaks down concepts into the individual pieces to be developed by other specific teams (architecture, engineering, etc.).

ARCHITECTURE translates concept illustrations and artwork into the various architectural drawings (plans, elevations, etc.) needed to turn the Imagineers' vision into reality.

ENGINEERING disciplines focus on the more technical aspects of an attraction, such as the show and ride control systems that make the attraction function. Specific types of engineering involved in developing these systems include:

- Mechanical Engineering: mechanical systems such as automatic doors, bridges, and structures.
- Electrical/Electronic Engineering: electrical and electronic systems that enable the attraction to operate.
- Ride and Ride Control Engineering: ride vehicles and systems, as well as how ride vehicles move and are controlled as they make their way through an attraction. A

great example of complex ride control engineering is in the Twilight Zone Tower of Terror at Hollywood Studios, where the ride vehicle not only moves up and down in the elevator shaft, but also travels along a trackless system through various show scenes.

- Audio-Animatronics Engineering: if an attraction includes animated figures, Audio-Animatronics engineers are responsible for their mechanical design and programming their movements, as with the birds in Walt Disney's Enchanted Tiki Room; Abraham Lincoln from Great Moments with Mr. Lincoln on Main Street, U.S.A. in Disneyland; the U.S. presidents in the Hall of Presidents at Magic Kingdom; and the Shaman of Songs in the Na'vi River Journey in Pandora at Animal Kingdom.

- Show Control Engineering: control and choreography of various show elements, such as Audio-Animatronics, video and audio, and the ride system. For example, as the ride vehicle enters different show areas in Journey into Imagination with Figment, it triggers different video and audio clips and other show effects. Perhaps one of the best examples of show control engineering is the American Adventure at Epcot, in which the various show scenes (all of which include Audio-Animatronics) are moved into place via an enormous carriage located below the stage and audience while scene-specific video and audio are channeled through different projectors and speakers (for more details about the show control systems in American Adventure, check out *Building A Better Mouse* by Steve Alcorn and David Green).

- Industrial Engineering: the overall layout of areas that guests will move through, including the layout and arrangement of the walkways and doorways that lead into and out of attractions. One specific area of industrial design focus is the queue. The size and length of an attraction's queue, including both outdoor queue space and indoor queue space, is largely determined by its expected capacity (which in turn is based on the attraction's Theoretical Hourly Ride Capacity [THRC], or "the number of guests per hour that can experience an attraction under optimal conditions").

SHOW WRITING develops scripts, spiels, and nomenclature for attractions, as well as any required text for signs and placards.

COSTUME DESIGN dresses the various characters in the attraction appropriately, much like theatrical costume designers dress the characters in a stage show. A great example of costume design in the Disney parks is the Hall of Presidents at Magic Kingdom, where all of the clothing worn by the Audio-Animatronics figures are authentic reproductions of the respective eras in which the presidents lived. For instance, the clothes on the Abraham Lincoln figure are authentic reproductions of the Civil War era.

LIGHTING DESIGN creates lighting effects for the show, as well as facility lighting for the building's exterior and interior lighting in the queue and the pre-show and post-show areas.

SOUND DESIGN is responsible for the sound effects heard during the attraction. It can also include sound effects intended to help establish a specific mood in the queue or pre-show. For example, several of the interactive features in the queue at the Haunted Mansion in Walt Disney World trigger specific sound effects. These are the domain of the sound designers.

EFFECTS DESIGN involves design of any special effects used in the attraction (the Pepper's Ghost illusion used in the Grand Ballroom of the Haunted Mansion being one of the best known).

MEDIA DESIGN develops and produces video or audio used in the attraction, *e.g.,* pre-show films, pre-recorded spiels, etc.

PROP DESIGN make the attraction look lived in and alive by designing the items and objects you see lying about its queue, show scenes, and pre-show and post-show areas.

INTERIOR/EXTERIOR DESIGN is responsible for design of the interior and exterior of the facility, including wall treatments (paint colors, wall paper, etc.), flooring (wood, carpeting, tile, etc.), doorways, and trim.

LANDSCAPE/HARDSCAPE DESIGN involves design of landscape in and around the attraction, as well as connected paths, walls, and other man-made features.

DIMENSIONAL DESIGN, or model making, involves designing and building three-dimensional models of entire attractions as well as individual show scenes, ride vehicles, and other aspects of the project to help work out design challenges.

GRAPHIC DESIGN produces marquees, flat and dimensional signs, and other artwork and patterns that help bring the vision to life. Examples of graphic design in the Disney parks include directional signs, attraction signs and marquees, and the unique wallpaper found in attractions such as the Haunted Mansion.

PRODUCTION DESIGN starts with the show design, takes it to the next level of detail, and ensures that it can be built so as to maintain the creative intent. It also must integrate the show with all the other systems that will need to be coordinated in the field during installation.

These descriptions barely scratch the surface of the varied Imagineering disciplines involved in the Design stage. For more information about these (and other disciplines), I recommend "The Imagineering Field Guide" series by Alex Wright and *Theme Park Design* by Steve Alcorn.

Drawings, Diagrams, and Documents

The output of the Design stage includes a wide variety of design documents and other objects, ranging from architectural blueprints to wiring schematics to story treatments and models. Let's look at some of the more common types of design documents the Imagineers develop during this stage.

One of the most fundamental architectural diagrams is the *plan*. As described in "The Imagineering Field Guide" series, a plan is a "direct overhead view of an object or space. Very useful in verifying relative sizes of elements and the flow of guests and show elements through an area." Two specific types of plans developed by the Imagineers are floor plans and site plans. A floor plan provides "a view from above showing the arrangement of spaces in [a] building in the same way as a map, but showing the arrangement at a particular level of a building" while a site plan "is a specific type of plan, showing the whole context of a building or group of buildings. A site plan shows property boundaries and means of access to the site, and nearby structures if they are relevant to the design."

Another important type of architectural drawing is the *elevation*. "The Imagineering Field Guide" series describes elevations as a "drawing of a true frontal view of an object—usually a building—often drawn from multiple sides, eliminating the perspective you would see in the real world, for clarity in the design and to lead construction activities." (The books in this series contain many examples of elevations.)

Another common type of architectural drawing is a *section* or cross section. "The Imagineering Field Guide" series defines a section as a "drawing that looks as if it's a slice through an object or space. This is very helpful in seeing how various elements interrelate. It is typically drawn as though it were an elevation, with heavier line weights defining where our imaginary cut would be."

Interior and exterior designers often create *colorboards* to communicate the specific types of colors and textures to be used for particular areas of an attraction. In *Designing Disney: Imagineering and the Art of the Show*, John Hench describes the Imagineers' use of colorboards in this way:

> We define our color selections first in storyboards, and then later in colorboards, which help us to select colors for any interior or exterior detail. In some cases, we include color sketches of the interior or exterior views of a structure as an additional guide.

Lighting designers communicate details of the lighting design by creating *light plots*, defined as:

> A light plot, lighting plot, or just plot is a document similar to an architectural blueprint used specifically by theatrical lighting designers to illustrate and communicate the lighting design to the director, other designers and finally the master electrician and electrics crew. The light plot specifies how each lighting instrument should be hung, focused, colored, and connected. Typically the light plot is supplemented by other paperwork such as the channel hookup or instrument schedule.

Sound designers create *sound design documents* to outline the various ambient sounds, music, and dialog guests will hear as they approach and experience an attraction.

Show writers create a number of specific types of written documents, including scripts and Show Information Guides. Like the script of a movie or television show, an attraction's *script* outlines the "action" performed by the characters in a given scene, as well as any dialog spoken by those characters. Memorable dialog from Disney theme park scripts include the Ghost Host at the Haunted Mansion who tells us "your cadaverous pallor betrays an aura of foreboding" and the talking skull in Pirates of the Caribbean who reminds us that "Dead men tell no tales!" According to the Imagineering master glossary in *Walt Disney Imagineering*, Show Information Guides are "created by Imagineers so that cast members have a detailed reference to the ride, show, restaurant, etc. The SIG contains the story, significant details, script summary, and spiels."

Though not technically documents, *models* are an important product of the Design stage. Precise scale models of an attraction and its key elements are an important tool during the Construction stage of the process. We'll look at models in more detail in Chapter 9.

Beyond these, other types of design documents produced during the Design stage include detail drawings, engineering drawings, and specifications:

- **DETAIL DRAWINGS** show a small part of the construction at a larger scale, to show how the component parts fit together. They are also used to show small surface details, like decorative elements.

- An **ENGINEERING DRAWING**, a type of technical drawing, is used to fully and clearly define requirements for engineered items.

- **SPECIFICATIONS** are written requirements for materials, equipment, systems, standards and workmanship. They contain what can't be communicated by a drawing.

In the early days of Imagineering, all of this design work was done on paper, but more recently most design work is done using computer software that allows various designs to be brought together into a single model through a process known as Building Information Modeling (or BIM). We'll look at this in more detail in Chapter 9 as well.

Milestones: 30%, 60%, and 90%

As the Imagineers work through the Design stage and as the amount of detail and information about the project increases, they employ a handful of milestones, or checkpoints. These checkpoints happen at approximately 30%, 60%, and 90% of the completed and finished design, allowing the Imagineers working on different aspects of the design to look at each other's work and identify any potential conflicts or gaps, and to make sure the overall design effort is staying true to the project's creative intent. In addition, they allow outside stakeholders (such as operations, merchandise, food & beverage, etc.) to provide feedback on the design.

From Concept to Reality

The goal of the Design stage is to *develop plans and documents that describe and explain how your vision will be brought to life.* The specific means by which this is done can vary greatly from project to project. Like was the case with Concept Development, the size, scale, and scope of your project will determine the extent of your work in the Design stage.

The Design stage practiced by the Imagineers works on two levels: the macro level (Facility, Ride, and Show design) and the micro level (individual disciplines and designers working within each of the three areas). This macro-micro level distinction isn't unique to Imagineering. The same can also apply to projects beyond the berm. As you consider the design of your project, what are the main areas you need to focus on? Are there parallels to Facility, Ride, and Show Design for your project? For large projects, the macro level might be major parts of the project. For example, the Apollo Lunar Module (or LM) comprised two main pieces: the Descent stage, designed to support the landing and serve as a launch pad, and the Ascent stage, which served as the crew compartment and was launched from the surface of the Moon to rendezvous with the Command/Service Module in lunar orbit. While the two stages obviously had to work together, each was a separate area of focus during the Preliminary Design and Detailed Design stages of the project (the equivalent of the Imagineering

Concept Development and Design stages), and the design of each required input from dozens of specialized disciplines.

If Concept Development is the stage where you learn what you don't know yet, the Design stage is where you put your own specific expertise and knowledge to work. For instance, an instructional designer creating a course about accounting software may not understand all of the details of accounting, but they're able to bring experience and expertise in instructional design to the project which allows them to translate the topics and subject matter into an effective instructional experience.

The Imagineers involve people from dozens and dozens of disciplines when designing their projects, and different types of creative projects require different types of expertise. Involving the right people with the right expertise is a key to bringing a vision from concept to reality. Consider the following examples:

Designing and building spacecraft such as the Apollo LM involves the following types of engineering expertise:

- Electrical Engineering
- Mechanical Engineering
- Software Engineering
- Propulsion Engineering
- Communications Engineering
- Systems Engineering
- Computer Engineering

Developing a Broadway musical involves the following types of creators and designers:

- Writer (responsible for the overall story, dramatic structure, and character development)
- Lyricist (responsible for writing song lyrics)
- Composer (responsible for the musical score)
- Choreographer
- Costume Designer
- Set Designer
- Lighting Designer
- Sound Designer

Designing video games involves the following types of designers:

- Level Designer
- Technical Designer
- System Designer
- User Interface (UI) Designer
- Tools Designer
- Sound Designer

In terms of the output of your Design stage, not all creative projects require extensive architectural drawings, specifications, and the other documents created by the Imagineers. Some may require only a handful of detailed outlines or storyboards (for example, a classroom lesson or training course), some might require hundreds of detailed engineering drawings and specifications (for example, electronic devices such as smart phones or tablets), some might call for dozens of detailed design documents (for example, software programs and/or video games), while others might require more unique creations (for example, the musical scores and scenic and costume designs needed for a theatrical production, or the different types of illustrations used in fashion design). The specifics of your project will determine the types of documents or specifications you'll need.

Lastly, as you progress through your Design stage, it's probably a good idea to have some checkpoints along the way to make sure everyone working on the project is in sync and that different pieces of the design will work together. This is useful mainly for projects with large teams. If you're doing all the design work yourself, you probably won't need to check in with yourself on your progress.

The Process in Practice

When writing this book, my Design stage was where I turned my outline and mind maps into working documents and where much of the actual writing took place. I created draft documents for each chapter using Scrivener (the tool I'm using to write this book), and compiled and organized my notes and research into a useful form. Within each chapter, I worked out the specific topics I wanted to address and how I would organize them, and also identified specific examples that would highlight each. This was also the stage where I did most of the actual writing.

Post-Show: Imagineering Checklist

- The goal of the Design stage is to *develop plans and documents that describe and explain how your vision will be brought to life.*
- Design can take place at the macro and micro level.
- Involving the right people with the right expertise is key to bringing a vision from concept to reality.
- The specifics of your project will determine the types of documents or specifications you'll need.

QUESTIONS

- What types of design documents do you need to build your project?
- What are the disciplines involved in designing your project?
- Are there macro and micro levels of design for your project?
- Are there parallels to Facility, Ride, and Show Design for your project?

CHAPTER EIGHT

Construction

The last major stage of the Imagineering Process is the Construction stage (also referred to as Implementation by WDI). It involves the actual physical construction and implementation of the project. This is where all the concept and design work from the previous stages comes together and the project takes the last step from concept to reality.

The work done at this stage includes physical construction of the facility, fabrication of specialized items such as props and ride vehicles, and installation of show elements to bring the attraction to life. In *Walt Disney Imagineering*, the Imagineers explain:

> It's time to put on our hard hats. Simultaneous efforts at the project sites and at Imagineering are in high gear. Foundations are poured and structural frames erected at the site while ride vehicles and animated figures and props are programmed and tested at Imagineering. Voice recordings, sound effects and

music are being mixed into the final digital soundtracks while a construction crew whips up a batch of plaster to trowel into the new facility.

At this point in our process, a show's support equipment, lighting fixtures, ride and audio systems—well, just about every piece of hardware—have been selected. Sets, props, and set dressings are hand-fabricated at the Imagineering production facility in North Hollywood.

Facility construction can range in size and scope from demolition and remodeling of an existing facility to full-scale ground-up construction including clearing and preparing the land, pouring foundations, erecting steelwork, running utility lines, framing buildings, etc. Just like the Design stage involved different types of designers, physical construction involves any number of construction workers, tradesmen, and craftsmen, including concrete finishers and cement masons, ironworkers, carpenters, electricians and plumbers, and interior and exterior painters.

As the attraction's facility begins to take shape, portions of the ride system, show and ride control systems, and show elements are installed as and where appropriate. For example, when the Imagineers were building Expedition Everest at Animal Kingdom, they installed portions of the ride track at or near the same time as the first pieces of structural steel were put into place since the ride track passes through portions of the mountain, and installing the ride track after the mountain was completed would have been impractical.

A common part of the Construction stage for many attractions—particularly those with a natural exterior such as Expedition Everest or Radiator Springs Racers at Disney California Adventure—is rockwork. The Imagineers explain:

> Creating outside show environments is not just limited to architecture and themed landscaping. Some of our finest shows have been created by simulating rocks and mountains. Re-creating a natural setting is difficult, creating a fantasy setting that must appear to be natural is even harder. That is where well-executed rockwork—a fine art in itself—comes into the picture.

Construction of rockwork begins with concept designs and detailed models and involves a process by which the models are scanned and digitized and used as the basis for "chips" fabricated from rebar and wire mesh. In a presentation about Imagineering for the Creative Mornings organization in Orlando Florida, Imagineer Jason Grandt describes the rockwork process:

> A lot of what we do is we build immersive environments and these things don't exist in reality, so a lot of times it involves creating landscapes. ... It all starts off with a scale model, and then it get scanned into a computer. The computer then breaks that surface up into individual chips, which another computer then takes rebar and bends it into chips, then labels them and numbers them, and then we build the thing. It's almost like a LEGO kit. The steel goes up, and then the chips are put into place, and then we start layering concrete.

Once the concrete is layered onto the rebar and wire mesh structure, it's then sculpted to form the desired shape of the attraction's surface and eventually painted and weathered as appropriate (practices known as "character plaster" and "character paint," which we'll look at in more detail below).

In addition to constructing the attraction's facility and rockwork, another key part of the Construction stage is the fabrication and production of props, vehicles, and other items. This is known as fabrication design, which, as described in the "Imagineering Field Guide" series, "involves developing and implementing the production strategies that allow us to build all the specialized items on the large and complex projects that we deliver." When you consider the wide range of props and other specialized items found in Disney theme park attractions, it becomes clear that fabrication is a major aspect of the Imagineering Process. Examples of fabrication design include:

- Ride vehicles such as the steam engine tea trains in Expedition Everest, the mine cars in the Seven Dwarfs Mine Train, the Time Rovers in DINOSAUR, and the Slinky Dog train in the Slinky Dog Dash in Toy Story Land
- Specialized ride systems and tracks, such as the rotating track used in Expedition Everest when the ride vehicle stops and then travels in reverse into the mountain
- Show scenes and sets such as the trees, fences, and buildings, as well as the Br'er Bear, Br'er Fox, and Br'er Rabbit Audio-Animatronics figures in Splash Mountain

- The crates and props that decorate the queue and the area around Big Thunder Mountain
- The gravestones, busts, and other items in the cemetery queue of the Haunted Mansion

Once these items are designed, built, and tested, they are delivered to the project site and installed as part of the overall show installation. As the Imagineers explain:

> Because most of our projects are so unusual, they contain show elements that are custom-made at either our Glendale or North Hollywood facility, sort of like making a ship in a bottle. Whether it's a large set, prop, oversized Audio-Animatronics figure, or ride vehicle, they are all designed and fabricated here at "home." Once they have been assembled and detailed, and meet the approval of the show team, the show elements are tested to make sure they perform as designed. Having passed this test, they are disassembled, packed up, and shipped for installation to one of our theme parks somewhere in the world. In most cases, an overnight delivery package just won't do.

The Construction stage is also where audio or video media used in the attraction is recorded, edited, and assembled for implementation in the attraction. In addition, any specific audio or video components—such as speakers or video monitors and displays—are built and installed into the attraction. Speakers and video displays are often built into the attraction's props and sets, such as rocks and landscaping, lighting poles, and other unusual places.

Show installation includes installation of the various show elements to the attraction, including sets, Audio-Animatronics figures, props, and other decorative elements. Show installation also involves interior and exterior decorating, including painting walls, applying wall paper, hanging artwork, and planting trees, flowers, and plants used in the landscaping in and around the attraction. As part of the show installation, two specific techniques used to help bring the attraction to life are Character Paint and Character Plaster. Imagineer Alex Wright describes these techniques in the "Imagineering Disciplines" sections of the "Imagineering Field Guide" series:

- **CHARACTER PAINT** "creates the reproductions of various materials, finishes, and stages of aging wherever we need to make something new look old." Character paint is how even when brand new, the façades of the buildings in the village of Serka Zong and the shrines in the Yeti temple at Expedition Everest look hundreds of years old.
- **CHARACTER PLASTER** "produces the hard finishes in the park that mimic other materials. This includes rockwork, themed paving, and architectural façades such as faux stone and plaster. They even use concrete to imitate wood!" Character Plaster is a key part of building the rockwork seen in so many Disney park attractions.

Test and Adjust

When the Imagineers finish the Construction stage, they move into a period called Test and Adjust, which the Imagineering Master Glossary defines as:

> The period in which an attraction is tested in the field before being turned over to Operations. The Test is running the show to make sure the THRC (Theoretical Hourly Ride Capacity) is met; Imagineers adjust things until we get the desired results.

While technically a separate stage in the Imagineer's process, for our simplified version of the Imagineering Process, Test and Adjust is considered part of the Construction stage since it takes place at the end of the construction effort.

At our lunch with an Imagineer back in 2011, Imagineer Jason Grandt shared a story about the Crush 'n' Gusher water slide at Typhoon Lagoon water park at Walt Disney World that serves as a good example of Test and Adjust in practice. The Walt Disney World website describes Crush 'n' Gusher as "a water coaster raft ride featuring 3 different fruit-themed waterslides." These three slides—Pineapple Plunger, Coconut Crusher, and Banana Blaster—were originally conceived and designed to accommodate two riders per raft, but during Test and Adjust, some of the lifeguards helping with the testing suggested changing this to allow three riders per raft. Testing this suggestion revealed that the Coconut Crusher and Pineapple Plunger slides could accommodate three riders, but the Banana Blaster could not. Based on this testing, the Imagineers made minor changes to the attraction to allow for the change in the number of riders allowed on the different slides. It was too late to make changes to the Banana Blaster slide to accommodate three riders per raft, but the Imagineers were able to change the attraction's signage to explain that the Coconut Crusher and Pineapple Plunger slides allowed two or three riders while the Banana Blaster allowed only two riders. This change also had an impact on the attraction's queue, since the queue was originally designed for an hourly ride capacity based on two riders per raft on each slide, and this change allowed for an increase in capacity of up to 30%.

Making It Real

The goal of the Construction stage is to *build the actual project, based on the design developed in the previous stages.* All of the concept and design work developed during the previous stages is brought to life in this stage. The specific activities involved in constructing your project will of course depend on the nature of project. For example, if you're creating a new training course, you may need to create presentations, write documents, or record and edit video and audio content; if you're remodeling a room in your house, you may need to build walls, install carpet, paint or hang wallpaper, etc.

For most projects, construction doesn't just mean assembling the final product, but also includes creating and fabricating the various pieces and parts used to assemble the project. Just like the Imagineers need to fabricate props, vehicles, and show elements when building Disney park attractions, you need to build the components that make up your project. For example:

- If you're creating a website, you would need to create the graphics, logos, and other visuals you intend to use on the site.
- If you're creating a board game, you would need to create the game board and pieces, as well as the game's rules and other components.
- If you're creating a video game, you would need to develop the game's engine, its user interface, models,

and backgrounds, as well as put it all together and play-test it.

- If you're creating a stage play or musical, you would need to construct the sets and props, create costumes, orchestrate and arrange the music, and conduct rehearsals.

Depending on the specifics of your project, this can involve building unique and custom parts and components. For instance, beyond standard nuts, bolts, and washers, most of the parts used in the assembly of the Apollo Lunar Module had to be designed and produced as part of the construction effort.

As was the case with the Design stage, this stage is where you put your own specific expertise and knowledge to work. If you're a graphic designer working on marketing or promotional materials, this stage is where you take your sketches and design notes and turn them into your final product. If you're an instructional designer, this is where you leverage your knowledge of learning styles and instructional techniques to combine lectures, hands-on activities, and other types of content to create an effective learning experience.

You'll also want to incorporate your own version of Test and Adjust as you assemble your project. As you build the parts and components of your project, each should be tested to ensure it meets your design goals and creative intent. Where items come up short or you encounter errors, be ready and willing to adjust things as needed. The same holds true for the final project. Testing of the individual parts and smaller components of your project along the way should help prevent major issues with your final project, but testing the finished item is just as important, if not more so. Remember that even the most exhaustive and comprehensive design effort leaves some room for error, and only thorough testing will help ensure your project is as close to perfect as possible.

In later chapters we'll look at examples of different types of creative projects, and each will include a look at the specifics of its Construction stage.

The Process in Practice

For this book, the Construction stage primarily involved the various tasks related to production of the final version of the book. This includes creating the final draft of the text (including the bibliography and other references), and then handing it over to the publisher for the creation of the cover design and interior diagrams, editing, layout of the book's text, and creating the final production-ready files for both print and electronic versions of the book.

Post-Show: Imagineering Checklist

- The goal of the Construction stage is to *build the project, based on the design developed in the previous stages.*
- Construction doesn't just mean assembling the final product, but also creating and fabricating the various pieces and parts used to assemble it.
- Incorporate your own version of Test and Adjust as you build your project.

QUESTIONS

- What are the specific tasks you need to complete to bring your idea to life?
- What parts, pieces, and components of your project can be built and tested prior to final assembly?
- Are you using your own version of Test and Adjust as you near completion of your project?

CHAPTER NINE

Models

As the Imagineers move from Blue Sky to Concept Development to Design to Construction, they use a variety of tools, most of which are specifically suited to particular stages of the process. In this chapter we're going to explore one of the tools the Imagineers use in all four main stages of the Imagineering Process: models. The Imagineers use models of different types and of varying sizes and scales to help identify and solve potential design challenges during the various stages of the Imagineering Process.

Models have been a foundational tool in the Imagineers' toolbox since the beginning of WED Enterprises, but the use of models in the Disney company goes back to the early days of the Disney animation studio, where animators and filmmakers used them to better understand and "see" the sets and scenes of their films. In a chapter about Disney animator and Imagineer Ken Anderson in *Walt Disney's Legends of the Imagineering and the Genesis of the Disney Theme Park*, Jeff Kurtti shares a specific example of the use of how Anderson used models during the production of *Snow White and Seven Dwarfs*:

> He [Ken Anderson] built scale models of the dwarfs' cottage to help the other animators visualize the settings dimensionally.

When Walt and his first Imagineers (who were artists, animators, and art directors from his animation and film studios) began their work on Disneyland, they adapted practices and techniques used in film making to three-dimensional storytelling and placemaking, and among these was the use of models. As Kurtti describes it:

> Walt loved to see things in an optimal reality. He knew from long experience at creating the "illusion of life" in animation that a painting or drawing could "fool" the viewer. He was more clearly able to see the possible problems, and potential enrichment, of a particular project by seeing it brought to life in three dimensions. So, when he began developing Disneyland, he founded the WED Model Shop. Walt is quoted as saying, "A model may cost $5,000, but it's sure less expensive than $50,000 to fix the real thing."

In *Walt and the Promise of Progress City*, Disney historian Sam Gennawey describes Walt's use of models:

> Walt loved using models for all of his ideas. Maybe it had something to do with his passion for miniatures. He was heard to say that drawings lie but models always tell the truth. Models can help the designer understand a project in new ways. A carefully crafted, highly detailed model can affirm the design direction or point out fatal flaws. This may explain why Walt required so many of his projects to go through modeling as part of the design process—even a project as large as a city.

Walt Disney Imagineering has continued this practice of using models, and in fact, it could be argued that the Imagineers currently use more and different types of models than at any time in the history of WED or WDI. In *Walt Disney Imagineering*, the Imagineers describe their use of models:

> In the development of a concept, the first models are simple ones fabricated from paper and foam blocks. Called massing models, their purpose is to introduce the idea to the dimensional world. These preliminary models demonstrate the

relationships of every aspect of the project to each other, and allow for discussion, analysis, and redesign as necessary.

As the project evolves, so too do the models that represent it. Once the project team is satisfied with the arrangements portrayed on massing models, small-scale detail-oriented study models are begun. These reflect the architectural styles and colors for the project.

Creating a larger overall model based on detailed architectural and engineering drawings is the last step in the model-making process. This show model is an exact replica of the project as it will be built, featuring the tiniest of details, including building exteriors, landscape, color schemes, the complete ride layout, vehicles, show sets, props, figures, and suggested lighting and graphics.

Put another way, as the design of an attraction evolves through the Blue Sky, Concept Development, and Design stages, the Imagineers create models of different types and scales which help them not only better understand how guests will experience the attraction and its environment, but also identify and solve potential problems in their designs. Beyond this, the Imagineers also use models in the field during the latter construction stages of their projects.

In *Designing Disney: Imagineering and the Art of the Show*, Imagineer John Hench describes the role of scale models:

At a later design stage, we use scale models, which are in a sense three-dimensional storyboards, to help us discover the relationships between time and space—the space through which guests will travel, the time it will take them to do so— within each element of the park.

With each iteration of an attraction's design, models help refine the Imagineers' ideas and vision. As Jason Surrell writes in *The Disney Mountains*:

Just as Harriet Burns and Fred Joerger had once interpreted the renderings of Herb Ryman and John Hench, the Everest model makers began with the sketches of Chris Turner and his fellow concept designers and built small paper models that were then sculpted first as a 1/8th-inch-scale clay model and later a foam model.

And just as Tony Baxter had once gone through nine iterations of Big Thunder Mountain...the Everest team went through a full twenty-four different models before they settled on a final design.

Models also play a role in integrating the various aspects of an attraction's design, for example, by helping the Imagineers determine how the attraction's ride system fits into the physical facility. Models often also help solve problems with an attraction's ride system, show building, physical location, etc.

A great example of this comes from the development of the ride system used in Soarin' at Disney California Adventure and at Epcot. As described by Imagineer Alex Wright in *The Imagineering Field Guide to Epcot* and *The Imagineering Field Guide to Disney California Adventure*:

> The Soarin' team was searching for possibilities of how to convey the sense of flight to a large audience with an acceptable capacity [when] Imagineer Mark Sumner came up with the solution to the particularly vexing engineering challenge of the Soarin' ride while playing with an old Erector set. He created a working model at home over a weekend. ... The simple piece, operated by a hand crank and demonstrating an action remarkably similar to the eventual attraction, spurred a new direction in thinking and a significant R&D effort to turn this toy into a reality."

The Imagineers use models during the Construction stage in multiple ways. In the previous chapter, we looked at how the Imagineers used detailed show models when installing and creating rockwork used in outdoor environments such as the mountains of Expedition Everest and Radiator Springs Racers, and the exterior of Under the Sea~Journey of The Little Mermaid at Magic Kingdom. In addition, the Imagineers use models when the practical realities of construction make it difficult or impossible for them to be able to see the whole project as they work on it. Alex Wright describes how models were used in this way in the design and construction of the Tree of Life at Animal Kingdom in *The Imagineering Field Guide to Disney's Animal Kingdom*:

> The development of the look of the Tree of Life was initially accomplished through numerous concept drawings and paintings, and then through an exhaustive model-making phase. The team carved renditions of the Tree in foam—first small scale, then much larger—and then tested various paint

and artificial foliage treatments in the model. This detailed process is what allowed the team in the field to maintain continuity when working at real size, as they were unable to step back and see the Tree as a whole during construction due to the scaffolding that was covering most of the surface.

Other Imagineering Models

Beyond these traditional types of models, the Imagineers also use other types of models, including Building Information Modeling (BIM), Pre-Visualizations, and Prototypes. Let's look at each of these.

BUILDING INFORMATION MODELING

In Chapter 7, I mentioned that in the last several years most Imagineering design work has been done using computer software. In *The Imagineering Field Guide to Disney California Adventure*, Imagineer Alex Wright writes:

> We still tend to begin with pencil on paper, and we still build physical models for all the same reasons we have in the past. In recent years, however, our design development processes have been irrevocably altered and improved. Contemporary modeling and drafting applications allow us to build virtual models very quickly and to do most of our design in 3-D from the beginning, and new time-based visualization tools combine our project schedule with the building model so that we can see it all come together before we ever get out to the site.

The "contemporary modeling and drafting applications" Wright describes are applications used with Building Information Modeling. The Imagineering Master Glossary defines a Building Information Model, or BIM, as the practice of "[c]reating a structured database that describes a building, usually in three dimensions; may contain additional information, such as time to construct and cost." Autodesk, "a leader in 3-D design, engineering and entertainment software," describes Building Information Modeling as "an intelligent 3-D model-based process that gives architecture, engineering, and construction (AEC) professionals the insight and tools to more efficiently plan, design, construct, and manage buildings and infrastructure." As the Imagineers develop different design diagrams and drawings, such as plans, elevations, sections, and detailed engineering drawings, these designs are incorporated into the attraction's building information model until the BIM is a fully developed three-dimensional computer model of the attraction. The Imagineers can use this to view the attraction from any direction, including exterior and interior views, and can also use the model to move through the attraction, beginning with the queue and passing through the entire experience.

In 2010, NASA held an Information Technology Summit where Jack Blitch, vice president of Walt Disney Imagineering in Orlando, described the process the Imagineers use when they design and build attractions, with a specific emphasis on their use of technology and modeling software, and more specifically the use of Building Information Modeling. Blitch's presentation focused on a handful of specific areas in which WDI uses BIM in their design process. The first area was detecting clashes between different elements of an attraction's design. As Alex Wright describes it:

> With these tools, we are able to detect and resolve conflicts between various building systems or elements and plan for any logistical challenge that will be encountered during construction.

Blitch's presentation includes a demonstration of "clash detection" using the building information model of Under the Sea~Journey of The Little Mermaid at Magic Kingdom. In this demonstration, the model detects a piece of structural steel that extends through the rockwork exterior of the attraction. Detecting this during design allows the Imagineers to make necessary adjustments in the model. In the past, these types of clashes would often go undetected until the Imagineers were on sight during construction.

Other specific uses of the BIM that Blitch describes are related to what he calls the model's fourth, fifth, and sixth dimensions. According to Blitch, the fourth dimension is time: WDI uses the BIM to estimate construction time and visualize the construction progress in the model. This also helps them understand when specific physical components of a project need to be delivered and available for installation. The fifth dimension is cost: in addition to time estimates, they use the

BIM to estimate the project's running cost at various stages of construction. The last dimension Blitch describes is what WDI calls the sixth dimension, which involves operations and ongoing maintenance of the facility. When WDI turns over an attraction to park operations (something we'll talk about more in the next chapter), they also turn over the BIM. Through integration with an asset management system, the BIM can provide operators access to documentation, technical specifications, and other information about equipment used in an attraction's facility, such as compressors and roof-top air handlers used for heating, ventilation, and air conditioning (HVAC). Through these fourth, fifth, and sixth dimensions, the BIM is used not only in the design stages, but also through construction and into ongoing operations. See Appendix A: My Imagineering Library for more information about Jack Blitch's presentation.

PRE-VISUALIZATION

Pre-visualization, another type of model used by the Imagineers, is borrowed from the film industry, where it is used "to visualize complex scenes in a movie before filming." In the case of theme park attractions, the Imagineers use pre-visualization to view and "experience" an attraction while it's being designed, and adjust their designs as needed. As Alex Wright describes it in *The Imagineering Field Guide to Disney California Adventure*:

> [Pre-visualization makes] extensive use of computer-generated imagery to experience a space or an attraction in a virtual manner from a guest's perspective. We have created for ourselves large-format projection environments that allow us to ride the attraction, walk the site, and look around corners just as a guest would—all with the confines of our design studios.

Pre-visualization tools use much of the same information and data used in building information modeling, but in a slightly different way.

Some specific examples of attractions designed using pre-visualization include California Screamin' at Disney California Adventure, "one of the first roller coasters designed with the assistance of computer pre-visualization, which allowed the team to *experience* the ride before it was built on-site," and Cars Land, also at Disney California Adventure. Imagineer and author Melody Malmberg provides a more detailed explanation of how the Imagineers used pre-visualization in the design of Cars Land in *Walt Disney Imagineering*:

> Using state-of-the-art computer hardware and interactive three-dimensional models, the Creative Technology Group taps into the power of virtual reality (VR) to help Imagineers see, experience, and present a project well before construction documents are finalized.
>
> To "previsualize" Cars Land in Disney's California Adventure, Imagineers with experience in the game and film worlds married scans of 3-D models, architectural files, engineering data, drawings, and color keys with banks of sophisticated hardware. A virtual Cars Land was displayed in a VR helmet or projected onto a huge curved screen built from a dozen high-definition monitors. Sensors allow the virtual environment to react to an Imagineer who could wander through the 3-D simulation at will. That way, a group could virtually jump into a vehicle and navigate the complicated ride track for the unbuilt Radiator Springs Racers, critiquing the layout in a years-early preview as the physical, small-scale model is being perfected. Virtual models allow decision-makers to visit and interact with the project and give them a visceral sense of the attraction and the guest experience well before construction begins.

PROTOTYPES AND PLAYTESTING

The last type of model used by WDI that we're going to look at is prototypes, commonly defined as "early samples, models, or releases of a product built to test a concept or process." The Imagineers use prototypes when developing ride vehicles and interactive features and attractions, as they provide a way to test and iterate their designs during the Blue Sky, Concept Development, and Design stages of the Imagineering Process.

A good example of the Imagineers using prototypes in the design of an interactive attraction is featured in the "Designs and Models" installment of *The Science of Disney Imagineering*, a series of educational DVDs that "uses roller coasters, stunt shows, Audio-Animatronics technology, and other irresistible examples to break down scientific principles in a way that is accessible for all learners." In one segment, host Asa Kalama interviews Imagineer and interactive show producer Estefania Pickens about how prototypes and playesting were used during the design and development of Toy Story Midway Mania at Disney California Adventure and Disney's Hollywood Studios. Talking about the use of prototypes in design, Pickens says:

> When you're designing something you always want to start with a simple model, because you don't want to spend too much with things you know you're probably going to change.

According to Pickens, one of the main questions the Imagineers tackled during the design of Toy Story Midway Mania was how to design an "interactive attraction that's fun for the whole family. In this case of Toy Story Midway Mania we answered a lot of our questions with playtesting." As Kalama explains:

> [Playtesting is] one way we Imagineers experiment with design. You have people use a prototype of your design—a functioning model—to find out what does and doesn't work. You record a playtest so when something doesn't work, you can study the problem and redesign.

In this interview, Pickens describes how the Imagineers used prototypes and playtesting to develop Toy Story Midway Mania, starting with the first prototype of the ride vehicle and spring-action shooter: a pair of chairs and an empty cardboard paper towel roll. They used this rough and primitive prototype to work out many of the specifics of the spring-action shooter, such as how far the shooters should move side to side and up and down. The next prototype was a full-scale wooden mockup of the ride vehicle that the Imagineers used to learn about how park guests would experience the ride—what they would see and from what angle they would see it. The mockup and playtesting also helped determine where to place the shooter so that guests of all ages could enjoy the attraction.

The Imagineers also used playtesting to work out details of the shooter itself. As Pickens describes it, "We came up with lots of different concepts and designs for our shooter." They went through and playtested about four or five different prototypes. Some required users to pull a lever, others had users spin a wheel, some involve pushing a button or plunger. The design they settled on uses a spring and ball mechanism, like an old-fashioned pop-gun toy. Once they had settled on its basic design, the Imagineers used further playtesting to work out specific details of the shooter, such as how long was the string, was the string comfortable for everyone to use, etc.

Models Beyond the Berm

The goal of creating models and prototypes is to *test and validate your design at each stage to help solve and/or prevent problems that may arise during the design and construction process.* Different types of creative projects require different types of models, some of which are not what we think of when we use the word "models." In the context of the Imagineering Process, any tool or process that helps you evaluate and refine your design can be considered a model.

In some fields, traditional three-dimensional models are used in much the same way as the Imagineers use them. Architecture is one such field, as author Matthew Frederick explains in *101 Things I Learned in Architecture School*:

> Three-dimensional models—both material and electronic—can help you understand your project in new ways. The most useful model for designing is the building massing model—a quick material (clay, cardboard, foam, plastic, sheet meter, found objects, and so on) study by which you can easily compare and test design options under consideration.

As the design of a building changes and evolves, models of each iteration of the design help architects make design and structural decisions and refine their concepts. Well-known architect Frank Gerhy is a famous example of using architectural models to work through design ideas. According to author Peter Sims, "Gehry and his team would, in fact, create eighty-two prototype models" during the design process for the Walt Disney Concert Hall in Los Angeles.

A specific type of model used in some fields, particularly technical fields, is a mockup, "a scale or full-size model of a design or device, used for teaching, demonstration, design evaluation, promotion, and other purposes." When designing and building the Apollo Lunar Module (LM), the design team used three different mockups during their Preliminary Design and Detailed Design stages to uncover and resolve challenges in the spacecraft's design. As project engineer Thomas J. Kelly describes in *Moon Lander*:

We needed the mockups in order to fit check the LM's complex shapes and assemblies and to determine whether astronauts and ground crew would be able to perform all required functions during the mission and during pre-launch preparation and maintenance.

Prototypes are another useful type of model, especially during the Blue Sky and Concept Development stages. Perhaps the most important thing in creating a useful prototype is that it helps you understand an idea and whether that idea is worth pursuing. As Tim Brown, CEO and president of IDEO, explains in *Change by Design*, "anything tangible that lets us explore an idea, evaluate it, and push it forward is a prototype." Prototypes can take any number of forms, even forms that are quite different from what you are trying to design. For instance, a stack of colored index cards might be used to create a prototype of a software application. Consider also the example of the initial prototype of the ride vehicle and shooter for Toy Story Midway Mania: a pair of chairs and an empty cardboard paper towel roll.

The goal of creating prototypes is not to solve your design problems as much as it to continually learn about what works and what doesn't as you refine your ideas. As Brown explains:

> The goal of prototyping is not to create a working model. It is to give form to an idea to learn about its strengths and weaknesses and to identify new directions for the next generation of more detailed, more refined prototypes.

Because most prototypes have a physical and tangible form, we tend to think of them as being useful only for projects that have a physical form as well, such as a vehicle or gadget, but the truth is, prototyping is useful when working on all sorts of projects. Prototypes can also be used when designing processes, services, or other less tangible projects. As Brown notes:

> There is also a role for prototyping more abstract challenges, such as the design of new business strategies, new business offerings, and even new business organizations. Prototypes may bring an abstract idea to life in a way that a whole organization can understand and engage with.

A specific type of prototype, typically created near the start of the design process and often the very first prototype, is a "proof of concept," or POC, described in the Imagineering

Master Glossary, as "the prototype of an idea that proves that it will work. It can be theoretical or actual." POCs are often used in software development as a rough, initial test of whether a particular design approach will work. For instance, Adobe produces a number of software application suites, including Adobe Creative Suite and Adobe Technical Communication Suite. These suites comprise sets of applications intended to be used together that were originally designed to be used separately. These suites started out as proofs of concept used to determine if and how the applications could be integrated into a single suite. Once a POC has been successful, it typically leads to additional prototypes which eventually lead to working models.

Most of the types of models we've looked at so far are what we tend to think of when we use the term "model," but some creative fields use different tools and approaches that serve as models to test and validate design decisions. For example, in the world of fashion and costume design, designers often use test garments to experiment with their designs before committing to creating formal patterns and cutting actual expensive fabrics. These garments are called muslins, described by Alfredo Cabrera with Matthew Frederick in *101 Things I Learned in Fashion School* as "a prototype garment, created to work out design and fit, called a muslin regardless of the fabric." Another example of a model-like tool comes from the field of music composition, in which composers use "musical sketches," a term defined by Friedemann Sallis as "the vast variety of documents that are used by composers to work out a musical technique or idea and to prepare their work for performance or publication." While the image of famous composers such as Beethoven or Mozart scribbling on pieces of parchment may come to mind, musical sketches are not limited to specific formats. As Sallis explains:

> Composers have used all kinds of objects and devices: from the erasable *cartella* of the sixteenth and seventeenth centuries to the digital screens of the late twentieth century.

Another less-than-traditional type of model can be found in the theater world, where new shows are developed and fleshed out through a process known as "workshopping." In a *New*

York Times article dated June 20, 2004, entitled "THEATER; Workshopped To Death," Jesse McKinley explains:

> [Workshops came into prominence] at least from an artistic standpoint...in the mid 1970s, when director Michael Bennett did a workshop on a musical and ended up with a classic: *A Chorus Line*, which ran for nearly 15 years on Broadway. Suddenly, workshopping seemed the key to building the perfect show.

Theater producers use workshops to work out a show's key design elements, including the show's book (its story or narrative structure), its songs, and even its choreography and staging. As McKinley writes:

> What is called a workshop can be anything from short and sweet—an afternoon's rehearsal and a quick reading by actors seated around a table—to long and grueling, involving weeks of rehearsal, daily rewrites and advice from everyone from dramaturge to audience members.

Other tools that function as models include storyboards, mind maps, and outlines. In *The Imagineering Pyramid*, I described storyboards in the context of filmmaking and theme park design as:

> ...large pin-up boards used to post ideas, or to outline the story points of a ride or film. Each story point or idea is on an individual sheet of paper or card. These allow designers to see the entire sequence of events in a story or ride, and re-arrange them as needed during development.

Any format or tool that helps you step back and see the larger picture of what you're creating can be thought of as a storyboard and can function like a model in your design process. For example, in instructional design you can use storyboards to outline the entire classroom experience (including lectures, quizzes, and exercises) and as a visual tool to "see" the entire course. Mind mapping is a practice that employs diagrams to organize and display information in a visual manner. Both storyboarding and mind mapping can be done with physical components or with computer-based tools. Lastly, traditional outlines (which in some respects are precursors to storyboards and mind maps) can also serve as models as you work through your design process. This is especially true in the case of written projects or information-based projects. In the case of writing projects, different drafts, such as a rough draft or first draft, can serve as models as a writer refines their story or subject matter.

The Process in Practice

I used a number of models and prototypes when writing this book, including mind maps and iterative drafts. I like to use mind maps for outlines, as they help me see how the chapters flow better than a text outline. In addition, when I ran into problems with specific chapters, I used mind maps to capture the main topics I wanted to address and played with different orders of topics until I found the right solution.

I went through three drafts of the book before sending it to the publisher. I tweaked my "in progress" draft constantly as I made my way from my Introduction to Post-Show. I tend to write from beginning to end, skipping around only in rare instances, but I often go back to previously finished chapters and make changes and alterations inspired by later chapters. After completing my first complete draft, I set the manuscript aside for a couple of days before reading through it, fixing errors and marking changes along the way. Once I had a second draft complete, again I waited a couple of days before reading through it again, this time reading it out loud to myself. It's surprising how many incorrect words, odd phrases, and other issues come up via this process. Finally, once I had polished the draft as best I could, I turned it over to my editor, who went through the manuscript himself, tightening up the text and correcting any outstanding issues I missed.

Post-Show: Imagineering Checklist

- The goal of creating models and prototypes is to *test and validate your design at each stage to help solve and/or prevent problems that may arise during the design and construction process.*

- Any tool or process that helps you evaluate and refine your design can be considered a model.

QUESTIONS

- How can you use models at the various stages of the Imagineering Process?
- Are you using playtesting to better understand how your audience will experience your project?
- Are you using prototypes to refine your ideas?
- Are there specific types of models you can use that are applicable to your field?
- Could you use a form of workshopping to refine and develop your project?
- Can you use storyboards to see the big picture of your project?

CHAPTER TEN

Epilogue: Openings, Evaluations, and Show Quality Standards

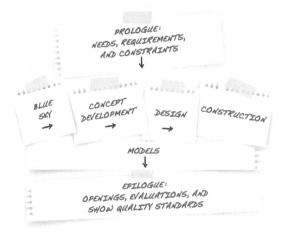

When construction and implementation of an attraction are complete, the process moves into the Epilogue, which comprises three parts: Openings, Evaluations, and Show Quality Standards. This is where the attraction is at last experienced by guests, and evaluated and maintained over time.

The event that signals the transition from Construction to Epilogue is the turnover or delivery of the attraction from Walt Disney Imagineering to the park operations team. This handover includes training cast members how to safely operate the attraction (including safety procedures such as evacuations) how to deal with show and ride malfunctions and stoppages, and what to do when the attraction goes "101" (code for a break-down that forces a closure). WDI also turns over operating manuals, maintenance manuals, Show Information Guides (we learned about these in Chapter 7: Design), and scripts. They also provide operations with the attraction's Building Information Model (described in the previous chapter), which provides access to documentation, technical specifications, and other information about equipment used in an attraction's facility.

Once cast member training is complete, most attractions go through a series of previews and openings in which larger and larger groups of people are allowed to experience it. These allow cast member operators to gain more experience in working on the attraction before it's opened to the general public. Attraction previews can include:

- Cast Member Previews (including Imagineers who didn't work on the project)
- Press/Media Previews (including traditional press, blog-gers, and podcasters)
- Disney Vacation Club Member Previews
- Annual Passholder Previews

Many of these previews are scheduled such as to limit the numbers of guests experiencing the attraction at once. In addition to the promotion these events bring to the attraction, they also provide park operations cast members additional training time with smaller crowds under (some-what) controlled conditions.

Once an attraction has begun these previews, it may also enjoy "soft openings," described in the Imagineering Master Glossary as when "operations cast members practice running an attraction before the official opening day, but with ordinary guests as well as their own family and friends. Soft Openings are not advertised." Not all attractions have soft openings, and a soft opening's days and times are unpredictable (to guests, at least) and can change quickly. Guests may luck into a soft opening of a upcoming attraction in the morning, only to return a few hours later to find it closed.

Previews, press events, and soft openings are all lead-ups to the big day when the attraction has its grand opening. This is typically a large event with lots of press and invited guests in which executives from Walt Disney Imagineering, the Walt Disney Company, and the park in which the attraction resides address the crowd before opening the attraction to park guests.

These same previews and openings are also used when opening a new land, such as was the case in May 2017 when Pandora opened at Animal Kingdom. May 2017 was a big month for WDI. While Pandora was opening at Disney World, across the country Guardians of the Galaxy: Mission BREAKOUT!, a "Re-Imagineering" of the Twilight Zone Tower of Terror, was opening at Disney California Adventure.

After an attraction has been open for some time, the park operations, management, and maintenance teams evaluate its success to determine how well (if at all) the attraction is meeting the needs and requirements that gave it birth. For example, if an attraction is built primarily to provide additional guest capacity, frequent break downs or slow guest flow could indicate that ultimately it didn't achieve its operational goals. These evaluations can include general guest satisfaction as well as specific operational evaluations, such as:

- How long is the attraction's "fill & spill"(the time it takes to load and unload a theater or show)?
- How long is the average stand-by wait time?
- How long is the average FastPass wait time?
- How often has the attraction broken down?
- What is the attraction's actual hourly capacity (as opposed to its Theoretical Hourly Ride Capacity, or THRC)
- How does the attraction handle its expected capacity?

Many of these evaluations are done through observation. For instance, park operations notes whenever an attraction "goes 101," logging not only the date and time of the event, but the duration and impact. They also track wait times throughout the day to better understand how guests flow through the attraction and nearby venues (such as a post-show gift shop). Other evaluations, such as guest satisfaction, are often done via in-park surveys.

The primary objective of these evaluations is to determine if, how, and how well an attraction meets its original

Need, Requirements, and Constraints. As we noted back in Chapter 4, every Imagineering project starts with a Need of some sort, and it is this Need that provides the business and creative motivation for the project. Should a project fall short in some way of meeting its original Needs or Requirements, or somehow violate one or more of its original Constraints, it could give rise to new Needs, Requirements, and Constraints and the potential of a whole new Imagineering project.

Beyond the park operations and management teams evaluating the parks and attractions, WDI also has a team of its own, the Show Quality Standards team, that continually evaluates the parks and attractions. The Imagineering Master Glossary describes Show Quality Standards (or SQS) as "a worldwide group of Imagineers who assure that the presentation in the parks maintain the creative intent of the original design. SQS team members inspect attractions and monitor new establishments in all parks."

SQS dates back to the early days of Disneyland and WED Enterprises, when Imagineers would work as supervising art directors for the park. This role was similar to what the SQS team does, as Jeff Kurtti explains in his essay about Imagineer Wathel Rogers in *Walt Disney's Legends of the Imagineering and the Genesis of the Disney Theme Park*:

> [Wathel Rogers] also served as art director for Magic Kingdom park after its opening and was the forerunner for what later became known as Show Quality Standards. Marty Sklar once remarked, "Any time a problem popped up in Walt Disney World, they would call Wathel and he always had a solution. They really loved him."

Imagineer Alex Wright elaborates on the role of SQS in *The Imagineering Field Guide to Magic Kingdom at Walt Disney World*:

> [The Show Quality Standards (SQS) program] is the means by which we ensure that the parks maintain their original design intent as they age, grow, and evolve over time. SQS groups work closely with park operators and maintenance teams to make certain that when any piece of the park needs to be replaced, it's put back as good or better. In addition, SQS reviews new designs to align them with existing park storylines and settings.

The outcome of both evaluations and Show Quality Standards are specific proposed updates and/or changes to be made to the parks and to specific attractions. These can range from short and simple refurbishments in which an attraction is freshened up (new paint, minor repairs, etc.), to more significant refurbishments including major repairs and restorations, to major redesigns and replacements such as the recent change at Disney California Adventure where the Twilight Zone Tower of Terror was replaced with Guardians of the Galaxy: MISSION BREAKOUT!, to the permanent removal of lands and attractions such as the recent closings of the Studio Backlot Tour, Lights, Motor, Action! Extreme Stunt Show, and most of the Streets of America at Hollywood Studios to make room for the upcoming Toy Story Land and Galaxy's Edge.

Changes and updates proposed by SQS can be based on a variety of different concerns, including:

- Artistic and aesthetic issues, such as faded or chipped paint on walls and doors and visibly worn ride vehicles
- Operational needs, such as a need to increase capacity and reduce wait times for particularly popular attractions.
- Potential safety issues, such as fatigued or loosened passenger restraints (these are exceedingly rare, given Disney's focus on safety)
- Routine maintenance, including regularly scheduled short refurbishments intended to keep the attractions in "good show" condition

Short and simple refurbishments are common; it's not unusual for various attractions to close for a few weeks for minor repairs, maintenance, some fresh paint, and a bit of polish. In some cases, short refurbishments also include minor changes to various show elements or scripts or spiels. Operational updates often result in longer projects since they typically include some amount of construction or change to the attraction's facility. Some specific examples of operational updates include recent changes to Soarin' at Epcot and Toy Story Midway Mania at Hollywood Studios. Both of these attractions often experience extremely long stand-by wait times, so both were expanded in 2016. In the case of Soarin' a third theater was added, while Toy Story Midway Mania got a third track installed. These updates increased the capacity of these attractions by roughly 50%, and have led to a significant reduction in the average stand-by wait time for each.

Often, these updates also provide the Imagineers an opportunity to "plus" the parks and attractions. Alex Wright defines Plussing as a "term derived from Walt's penchant for always trying to make an idea better. Imagineers are continually trying to plus their work, even after it's 'finished.'" The Imagineers use Plussing as a way to improve every element of the guests' experience, including updates and enhancements to not only the attractions themselves, but also attraction queues and surrounding areas. Some noteworthy examples of Plussing include:

- **STAR TOURS: THE ADVENTURE CONTINUES**. Originally this attraction featured a single story, but now it has randomized story sequences with film footage from all of the Star Wars films. In addition, the quality of the video has been dramatically enhanced with the use of high-def 3-D.

- **THE TWILIGHT ZONE TOWER OF TERROR** (at Hollywood Studios). The elevator drop was originally a simple pre-defined drop sequence which was the same every time, but it now features a randomized drop sequence so it's "never the same fear twice."

- **TEST TRACK**. In 2012, this attraction was significantly changed from its original concept of a General Motors testing plant to a Chevrolet Design Center where guests can design and test their own SIM Cars, with an entirely new design aesthetic.

- **UNDER THE SEA ~ JOURNEY OF THE LITTLE MERMAID**. An update included new lighting and painting that brought a classic Disney dark ride feel to key show scenes.

- **HAUNTED MANSION** (at Magic Kingdom). Updates to this classic attraction over the last several years have included new show scenes featuring Escher-like stairways, attic scenes featuring the Bride and her many deceased grooms, and interactive animated hitchhiking ghosts.

In addition, the queues of several attractions at both Disneyland and the Magic Kingdom have been plussed with interactive elements to help engage the audience and keep them occupied while they wait in line. Some specific attractions at Magic Kingdom that feature updated interactive queues include the Many Adventures of Winnie the Pooh, Haunted Mansion, Big Thunder Mountain Railroad, and Peter Pan's Flight.

Regardless of the size, scope, or type of update, changes born from evaluations or proposed by SQS form the basis of new Needs, Requirements, and Constraints, and often lead to new Imagineering projects. This brings the Imagineers full circle from the Epilogue of one project to the Prologue of another.

Sharing Your Ideas With the World

Once you've completed the Construction stage on your project, it's time to share it with your audience, whoever that audience might be. Audiences can take many forms depending on the type of project. If your project is a product you plan to sell, your audience might be your customers; in the case of a business-related project, it might be other stakeholders or management; if the project is a school assignment, that audience might be the teacher or other students. Regardless of your specific audience, the goal of the Epilogue is to *present your project to your audience, allow them to experience it, and evaluate its success and effectiveness over time.*

When presenting your project, you might present it to a series of larger and larger audiences, such as how new Imagineering projects are opened with different types of previews and openings. In business settings, for example, it's common for new policies or systems to be rolled out to larger and larger groups over time rather than presenting the project to the entire organization. For some projects these previews and limited openings provide the opportunity for a final round of "Test and Adjust" where minor adjustments can be made based on feedback from your preview audience. For example, in the area of training, it's not uncommon for training courses to undergo minor tweaks and changes following their "first teach" (the first time they are delivered to a fresh audience).

Once your audience has had the chance to experience your project, you should evaluate how well the project is received and its success. Part of this evaluation might be largely subjective (whether your audience liked it or not), but a part should also be based on more objective criteria, including the project's original Needs, Requirements, and Constraints. In Chapter 4, I offered some questions to help identify your Needs, Requirements, and Constraints. A little rewording of these provides questions you can use to evaluate your project, such as:

- Does your project solve the problem you designed it to solve? (Does it meet the original Need?)
- Does your project do all of the things it was supposed to do? (Does it meet the original Requirements?)
- Does your project do any of the things it was *not* supposed to do? (Does it violate any of the original Constraints?)

Another way to evaluate your projects is to use questions based on the principles from the Imagineering Pyramid. Did you:

- Use your subject matter to inform all decisions about your project? (It All Begins with a Story)
- Stay focused on your objective? (Creative Intent)
- Pay attention to every detail? (Attention to Detail)
- Use appropriate details to strengthen your story or subject matter? (Theming)
- Organize your message to lead your audience from the general to the specific? (Long, Medium, and Close Shots)
- Attract your audience's attention and capture their interest? (Wienies)
- Make changes as smooth and seamless as possible? (Transitions)
- Focus on the big picture? (Storyboards)
- Introduce and reinforce your story to help your audience get and stay engaged? (Pre-Shows and Post-Shows)
- Use the illusion of size to help communicate your message? (Forced Perspective)
- Simplify complex subjects? ("Read"-ability)

- Keep the experience dynamic and active? (Kinetics)
- Use repetition and reinforcement to make your audience's experience and your message memorable? (The "it's a small world" Effect)
- Involve and engage your audience? (Hidden Mickeys)
- Consistently ask "How Do I Make This Better?" (Plussing)

Beyond these types of evaluations, you might also want to adopt your own form of Show Quality Standards to ensure that your projects continue to serve their original objectives over time. In some cases, circumstances and audiences change and a project that originally met its objective might no longer do so. Continually evaluating your projects in the context in which they are experienced can help you identify if and when changes need to be made. For creative projects with an extended lifespan, it's especially important to occasionally evaluate whether or not they are still fulfilling their initial objective. For example, if you develop a system or process in your work to help solve a problem or a challenge, once the problem or challenge has been solved you may find that it's time to retire that process or system. Sometimes we continue to do things long after they've served their purpose because those activities become "the way we've always done things." Continuously evaluating your projects is a good way to identify when older projects no longer serve their original needs.

The Process in Practice

For this book, my previews and openings included me sharing final drafts of the book with colleagues and associates, and the formal release of the book by my publisher, Theme Park Press. Evaluations and feedback will come over time as people read the book and share their comments with me and (hopefully) post online reviews. This feedback could eventually lead to an updated or revised edition, or potentially an entirely new book in the Imagineering Toolbox series.

Post-Show: Imagineering Checklist

- The goal of the Epilogue is to *present your project to your audience and allow them to experience it.*

- Evaluations of a project's success can be both subjective and objective.
- You can use your project's original Needs, Requirements, and Constraints to evaluate how well you've achieved your objective.
- Continually evaluating your projects in the context in which they are experienced can help you identify if and when changes need to be made.

QUESTIONS

- How will you share your project with your audience?
- Can you use previews or soft openings to introduce your project to select audiences?
- How are you going to evaluate the success of your project?
- Do you have an equivalent of Show Quality Standards?

....

PART THREE

The Imagineering Process Beyond the Berm

In Part Two we looked at the individual stages of the Imagineering Process and how they are used by the Imagineers in the Disney parks, as well as how each can be applied "beyond the berm." In Part Three, we're going to look at how the Imagineering Process can be applied to some specific creative fields outside of Disney theme parks, including game design (Chapter 13), instructional design (Chapter 14), and management and leadership (Chapter 15).

Before we get to those specific fields, however, I want to first step back and look at the process as a whole again, explore its flexibility and adaptability, and consider some examples of how this process can be applied to simple types of projects (Chapter 11). I also want to look at examples of how the stages of the Imagineering Process align with the development of some specific creative projects (Chapter 12).

The chapters in this section are intended to provide examples of how the Imagineering Process can be applied "beyond the berm." Even if you don't currently work in any of these fields (in fact, I suspect that's the case for many of you), the examples of how the Imagineering Process can be used in these fields will hopefully provide some insight into how it can also be applied to your own field, whatever it may be.

CHAPTER ELEVEN

Another View of the Imagineering Process

In Part Two we looked at each of the stages of the Imagineering Process in detail, and it's easy to get so caught up in the specifics of each stage that we forget to step back and remember that the individual stages are part of a larger whole, and that the intent of each stage is to move your creative project one step closer from concept to reality. In this chapter, we begin to move out of the parks and explore how the Imagineering Process can be used with creative projects that lie beyond the berm. I want to start by taking another look at the process as a whole to see how the process really works, and how the stages work together. I also want to explore some aspects of the process that might not be so apparent up to now, as well as look at some simple and small-scale example projects.

Let's start by revisiting the diagram we've seen in previous chapters that outlines the stages of the Imagineering Process. The Prologue leads to the five core stages– Blue Sky, Concept Development, Design, Construction, and Models—which in turn lead to the Epilogue. And, as we saw in Chapter 10, the evaluations and Show Quality Standards in the Epilogue often bring us back to the Prologue with new Needs, Requirements, and Constraints.

In previous chapters we drilled down into each of the blocks in the process to better understand how each works on its own. But if we strip away the details of each stage and boil each down to its core essence or objective, we find a simple but powerful process suitable for nearly any type of creative project.

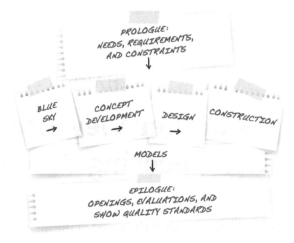

To bring your creative ideas to life:

- Define your overall objective, including what you can do, can't do, and must do when developing and building your project. (Prologue)
- Create a vision with enough detail to be able to explain, present, and sell it to others. (Blue Sky)
- Develop and flesh out your vision with enough additional detail to explain what needs to be designed and built. (Concept Development)
- Develop the plans and documents that describe and explain how your vision will be brought to life. (Design)
- Build the actual project, based on the design developed in the previous stages. (Construction)
- Test and validate your design at each stage to help solve and/or prevent problems that may arise during the design and construction process. (Models)
- Present your project to your audience, allow them to experience it, and evaluate its success and effectiveness over time. (Epilogue)

Flexibility, Scalability, and Adaptability: All Projects Are NOT Created Equal

Our look at the stages of the Imagineering Process up to now may have implied that it is somewhat rigid and that all projects move from stage to stage to stage in the same way. That's not the case at all. Because of its simplicity, the Imagineering Process can be used in different ways depending on the specifics of each project. Let's look at how the process, particularly the four core stages (Blue Sky, Concept Development, Design, and Construction) can be adapted to different types of projects.

At first glance, the core stages of the Imagineering process seem straightforward enough. We go from Blue Sky to Concept Development to Design to Construction.

However, life is rarely that simple. Any process used to bring creative ideas to life needs to be flexible, scalable, and adaptable. Fortunately, the Imagineering Process is all of those.

FLEXIBILITY AND ITERATION

It is unusual for the Imagineers to simply follow this process strictly stage-to-stage from Blue Sky through to Construction. More often, they end up using an iterative approach, in which they re-visit certain stages along the way as the needs of the project dictate. For example, while in the Concept Development stage, a problem may arise that causes the team to return to the Blue Sky stage to identify solutions to the new problem. Later on in the Design stage, it might become necessary to further develop some of the conceptual work done in Concept Development, or a new Need or Requirement is uncovered, sending the team back to Blue Sky. Perhaps a more accurate depiction of the process looks like this:

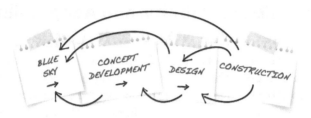

Imagineer Jason Grandt shared a story with me that provides a good example of this type of iteration. At one point during the construction and show installation of Town Square Theater at Magic Kingdom, Jason (who was serving as art director for the project) and other Imagineers on the team spent the bulk of one day looking at the various doorways in the attraction, including the doorways into the main lobby, the doorways from the lobby into the halls leading to Mickey's dressing room, and the doorways from the halls into the dressing rooms themselves. As a result, the team identified changes that needed to be made to various doorways in the attraction, and he and others worked on drawings of those changes later that same evening.

SCALABILITY: MACRO AND MICRO

The Imagineering Process is also flexible in that it can it works at both the macro level (the overall project), as well as the micro level (each small piece of the project).

The steps of the Imagineering Process outline the overall steps of an Imagineering project, but they can also apply to smaller sub-sections of the project as well. For example, as a project moves from Concept Development to Design, the various design teams, such as Lighting Design, Interior Design, Sound Design, Effects Design, etc., may each go through their own instance of the Imagineering Process as they conceptualize, develop, design, model, and build the various components of the project. Effectively what happens is that during Concept Development the Imagineers identify domain-specific Needs, Requirements, and Constraints, which form the basis for smaller scale versions of the process, and the overall Design stage is the amalgamation of a series of smaller-scale processes (each of which can be iterative, too!). This might be depicted as follows:

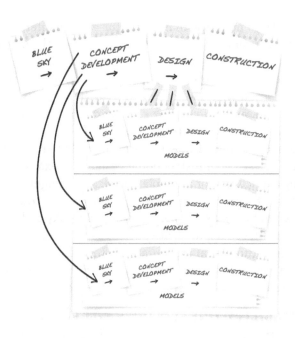

These macro and micro levels not only apply when designing new attractions, but also when the Imagineers are designing new lands or parks. The land or park is the macro level, while the individual attractions and venues are at the micro level.

For example, when designing New Fantasyland for Magic Kingdom, the overall project was initiated with a set of Needs, Requirements, and Constraints—expand the Magic Kingdom's capacity with a new area that fits thematically in Fantasyland. The project started in the Blue Sky stage and then moved into Concept Development where the Imagineers decided on the specific attractions and venues to be built in the new land, including Seven Dwarfs Mine Train, Be Our Guest restaurant, Enchanted Tales with Belle, and others. Each of these in turn became a full-scale Imagineering project of its own, with its own set of Needs, Requirements, and Constraints, starting with the Blue Sky stage and moving through the rest of the Imagineering Process.

This macro- and micro-level division applies to projects beyond the berm as well. For example, I went through a small version of the process when writing each chapter of this book.

Once I settled upon the main topic for each chapter, I had to develop a concept for how I could communicate the idea (Blue Sky), followed by research to ensure that I could support my ideas (Concept Development), fleshing out my outline in more detail so I knew what I was going to write (Design), and finally writing, editing, and rewriting the text until it made some semblance of sense (Construction).

Another example of how this macro- and micro-level division can be used is in instructional design. For example, when creating curriculum for a complex subject, you might begin the process at the macro level ("Create implementer training for our new product"). As you move through Blue Sky and Concept Development, you might identify a need to create more than one course, including separate courses for installation, configuration, end user, system administration, etc. Each of these courses in turn leads to a smaller scale ("micro") version of the process.

ADAPTABILITY: ONE SIZE DOES NOT FIT ALL

Along with being iterative and scalable, the Imagineering Process is also adaptable.

The process diagram we've used to this point implies that the four main stages (Blue Sky, Concept Development, Design, and Construction) are roughly equivalent in terms of size, scope, and effort. That is not necessarily the case.

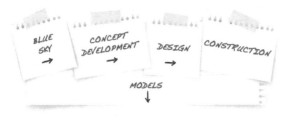

Generally speaking, Blue Sky and Concept Development are often roughly equal in terms of duration, though Concept Development often involves more people and a generally larger effort. The Design stage is typically longer in duration and involves even more people, while Construction is generally the longest stage in terms of duration and involves the most people and effort.

In practice, however, the various stages in the process can vary considerably in terms of level of effort and scope. For example, some projects may spend years in Blue Sky while others will move to Concept Development relatively quickly. Likewise, some projects don't require extensive Concept Development beyond the conceptual work created in the Blue Sky stage. With regard to the Design and Construction stages, smaller projects, such as meet-and-greets or new merchandise locations, will not likely require the same level of design or building effort as an entirely new attraction such as Expedition Everest or Seven Dwarfs Mine Train.

Let's look at some examples of how the stages of the process can vary in size and scope.

This first example is a project in which each stage is longer and more complex than the last. It starts with a small Blue Sky stage, with more time and effort spent in Concept Development, even more in Design, and a Construction stage that is longer and more involved that any of the previous stages:

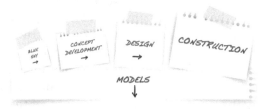

Another example is for a project with an extended Blue Sky stage followed by a slightly shorter Concept Development stage, a relatively small Design stage, and a larger Construction stage:

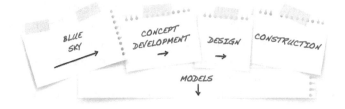

One last example has a short Blue Sky stage, followed by a longer Concept Development stage, an extended Design stage, and a shorter Construction stage.

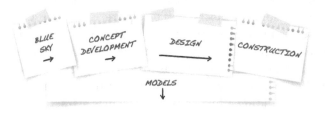

This variety occurs in all sorts of projects beyond the berm as well. For example, you might create a vision for your project (Blue Sky) quickly, but need to spend a considerable amount of time fleshing out your vision and developing your ideas before you understand what you really need to build (Concept Development). In some cases this extended Concept Development results in a relatively short and simple Design stage, but in others the Design stage could be just as extensive as Concept Development. Lastly, the Construction stage might be simple and straightforward while in others it might involve large-scale effort from lots of people. The specifics of each project will dictate the size and scope of each stage of the process. These examples are just a handful of the possible permutations of differently sized stages.

HOW MANY STEPS?

The last way in which the Imagineering Process is flexible and adaptable is related to the way a project moves from stage to stage through the process.

The basic process diagram clearly delineates the four main stages (Blue Sky, Concept Development, Design, and Construction), suggesting that there is always a clear separation between each stage. This implies a formality and rigidity that need not apply to every project.

When thinking about working on a small or simple project, you may question whether you need to go through all four stages. While each stage has separate goals and objectives,

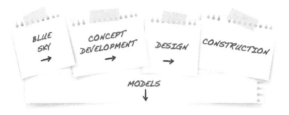

one of the primary goals of each stage is to prepare for the next stage, and in some cases you might find that the separation between the stages might not be so distinct as the basic process diagram implies. In fact, as we work on different types of creative projects, we may find that clear and distinct separations between stages do not always exist. Put another way, in some projects, adjacent stages may blur together somewhat such that the work in one stage bleeds over into the next. There are three variations of this.

The first is where the Blue Sky stage merges with Concept Development. This can happen when the Concept Development effort is effectively done as part of the initial concept design effort in the Blue Sky stage. This can occur with relatively simple projects, and looks like this:

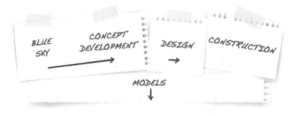

The second variation is where the Concept Development stage merges with the Design stage. This happens when fleshing out the vision during Concept Development results in enough detail that further design work is not needed. For example, when designing training materials for a simple product or feature, the outlines or mind maps created during Concept Development might be all you need to move on to the Construction stage. This variation looks like this:

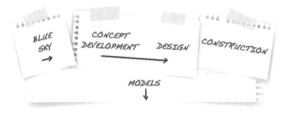

The last variation is where the Design stage merges with the Construction stage. This applies to projects in which the design effort is also a construction effort. One example of this is creating presentations. Very often designing a presentation also involves creating the presentation as well. Another example of this is writing and publishing a book. I believe that writing belongs in the Design stage, but it is also clearly part of the Construction stage as well, since you can't create a a book without its words. This last variation looks like this:

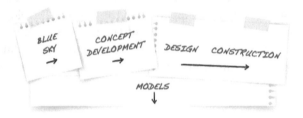

If we combine iteration and flexibility, scalability (macro/ micro), and variation in the size/scope of each stage as well as the number of stages, we see that the Imagineering Process offers an infinitely adaptable process suitable for nearly any type of creative project, inside or outside the berm.

Some Simple Examples

Now we've seen the stages of the process in detail, and understand how it all fits together. We even know that the process is flexible and adaptable, but what does a project look like as it moves through the process? How does an idea start as a Need and end up a reality? In the coming chapters we're going to look at some examples of the process in the context of some elaborate creative projects and some specific creative fields (game design, instructional design, and leadership and management), but for now, let's look at two small, simple examples of the Imagineering Process at work.

FUNDRAISING CAMPAIGN SLOGAN AND LOGO

In the winter of 2017, our local high school music department took a trip to Walt Disney World, during which the band marched in a parade in Magic Kingdom, and the school's a cappella groups performed on the Marketplace Stage at Disney Springs.

To help support this trip, our local music booster organization ran a number of fundraising events to provide as many opportunities as possible for the kids to earn money toward the cost of the trip. As president of the booster organization, I proposed that we wrap all of these events into a larger campaign that we could advertise around town to let folks know about the trip and the various fundraising events we'd be running. The goal was to create a slogan and logo for the campaign that would help us communicate our goals of bringing our high school music students to Walt Disney World. Let's look at how we used the Imagineering Process to develop this slogan and logo.

- **PROLOGUE**. The goal outlined above formed the Need for this project: "We need a slogan and logo for our fundraising campaign to help communicate what we're raising funds for." We had very few specific Requirements or Constraints for this project.
- **BLUE SKY**. We brainstormed ideas for the slogan and logo, experimented with a handful of ideas that we could choose from, and selected the final slogan ("Help Us Bring Our Music to the Mouse").
- **CONCEPT DEVELOPMENT**. We experimented with sample designs for the logo based on our selected slogan.
- **DESIGN**. We developed a final design of the logo (with the help of a parent who works as a graphic designer).
- **CONSTRUCTION**. Our graphic designer produced final art files that were used in our fundraising materials, such as raffle tickets, flyers, posters, website, etc.

- **MODELS**. As we worked through ideas in Blue Sky, Concept Development, and Design, we created review drafts and mockups and distributed them among the fundraising team for feedback and comments.
- **EPILOGUE**. We are still in the process of evaluating how helpful the slogan/logo were in promoting the fundraising efforts, and if we want to use a similar approach in the future.

ONLINE HELP AND USER DOCUMENTATION FOR A NEW PRODUCT

Another example comes from the work I do as a documentation manager at an enterprise software company. We recently released a new product formed by combining two separate existing products into a single application. Part of this project involved creating new user documentation and online help for the new application. Let's look at how we used the Imagineering Process in this case:

- **PROLOGUE**. Our Need was as follows: "We need to develop online help for a new product that combines functionality from two current products." In addition to this, we also had Requirements and Constraints mostly based on the deadline and the number of people we could devote to the project (very few in this case). We also wanted the customer experience for this new application to be that of a single application and not simply the combination of two existing products.

- **BLUE SKY**. We identified two main approaches for the new documentation: 1) using the standard help for existing products and adding new content as needed or 2) merging the existing documentation and editing the content as needed to create a single set of documentation for the new product. We selected the merging approach since it allowed us to portray this new application to users as a single application.

- **CONCEPT DEVELOPMENT**. We created an initial outline and mind map for the merged documentation, noting sections that required rearranging or deletion. We identified several new chapters and sections that needed to be written from scratch. We also created a proof-of-concept of merging files from different repositories to demonstrate to stakeholders that the merging approach was viable.

- **DESIGN**. We created a detailed outline listing every chapter, section, and topic in the merged documentation, and reviewed each in detail to identify which needed to be removed or edited. We also drafted detailed outlines for the new chapters or sections we identified during Concept Development.

- **CONSTRUCTION**. Working from our detailed outline, we updated the content for the new application, including writing new content as well as rearranging and editing existing content as needed. The final content was delivered as both context-sensitive online help, and as user guides in PDF.

- **MODELS**. At each stage we created artifacts and objects used to test and validate our design, including our original mind map outline, our detailed outline, and our proof-of-concept.

- **EPILOGUE**. As customers buy and implement this new product, we will solicit feedback to look for ways to improve the documentation for later releases.

CHAPTER TWELVE

A Trip to the Moon and Birthday Parties

Before we consider how the Imagineering Process can be applied to specific creative fields, I want to look at some specific creative projects that I believe are excellent examples of how the stages of the Imagineering Process work. These projects range in size, scope, and complexity from relatively small and simple to technically vast and complex. The first comes from the field of aerospace engineering, and the second is a game I created for my son's 8th birthday party.

Imagineering Apollo

The first example comes from the Apollo space program—specifically the design and development of the Apollo Lunar Module.

I was five years old when Apollo 11 landed on the moon and Neil Armstrong and Buzz Aldrin become the first men to walk on its surface. Like many kids growing up at the time, I was captivated by the space program and dreamed of becoming an astronaut. Also like many kids, my interest faded over time and I left my dream of flying into space behind. I remained casually interested in the space program as it moved on to Skylab, the Apollo-Soyuz mission, and eventually the Space Shuttle, watching shuttle launches on TV when I could and keeping tabs on various shuttle missions, including the Challenger disaster in 1986. But then my interest in the Apollo program was renewed when HBO aired the mini-series *From the Earth to the Moon* in 1998, a docudrama depicting the Apollo space program and moon landings. My wife and I watched the series when it first aired, and have re-watched it several times. It has become one of my all-time favorite TV series.

One of my favorite episodes is "Spider," which tells the story of the development of the Lunar Module and its first flight on Apollo 9. The episode is told from the point of view of Thomas Kelly, project engineer and program manager for the Apollo Lunar Module at Grumman Aircraft, the company NASA selected to design and develop the Lunar Module. Kelly also wrote a book about the project called *Moon Lander: How We Developed the Apollo Lunar Module*. The story of the development of the Lunar Module is a great example of creativity at work, and offers interesting examples of the various stages of the Imagineering Process. It is also an example of a creative project done under contract with a third party (in this case NASA) as opposed to a project done for more personal reasons.

Before I get into the specifics of the process and each stage, I want to provide some context about the Lunar Module and its role in the Apollo moon landings, since I think that will help frame the description of its development.

According to the Smithsonian National Air and Space Museum website:

> The Apollo Lunar Module was] a two-stage vehicle designed by Grumman to ferry two astronauts from lunar orbit to the lunar surface and back. The upper ascent stage consisted of a pressurized crew compartment, equipment areas, and an ascent rocket engine. The lower descent stage had the landing gear and contained the descent rocket engine and lunar surface experiments.

What role did the Lunar Module play in the Apollo lunar missions? Let's look at a simplified description of a "typical" Apollo moon landing mission.

The mission starts with the launch of the Saturn V rocket carrying three astronauts in the Command/Service Module (CSM), with the Lunar Module (LM) located in a compartment just under the CSM. After establishing orbit around the Earth, the third stage fires sending the rocket toward the moon (the first two stages do the job of getting the ships into orbit). On the way to the moon, the CSM separates from the Saturn V rocket and extracts the LM, and the rest of the rocket is abandoned. Upon reaching the moon, the spacecraft (both the CSM and LM) enters lunar obit. Two astronauts (the commander and the Lunar Module pilot) enter the LM, separate from the CSM, and descend to the lunar surface to land at the mission's pre-determined landing site. The astronauts on board exit the LM to explore the surface per the specific mission's objectives (the first moon walk lasted only two-and-a-half hours, while the longest lasted more than seven hours). Once the landing/exploration portion of the mission is complete, the astronauts launch the ascent stage of the LM and enter lunar orbit, for a rendezvous with the CSM. The commander and Lunar Module pilot transfer to the CSM along with the lunar samples they collected while on the moon. The CSM separates from the LM, leaving it in lunar orbit (where it eventually crashes on the lunar surface) and the CSM begins its journey back to Earth.

Now we that understand the Lunar Module and its role in the Apollo moon landing missions, let's look at its design and development through the lens of the Imagineering Process.

PROLOGUE: NEEDS, REQUIREMENTS, AND CONSTRAINTS

Define your overall objective, including what you can do, can't do, and must do when developing and building your project.

The Need, Requirements, and Constraints of the Lunar Module were provided in the form of a Request for Proposal (RFP) from NASA which outlined in detail the need for a vehicle that would support landing on the moon via "lunar orbit rendezvous," a process by which a small and separate vehicle is used solely for the moon landing. Beyond this basic need, some specific requirements included contractual, operational, and technical details, such as requirements around program management and facilities in addition to the technical requirements of the lunar module itself (it had to support a crew of two astronauts, be capable of landing and taking off from the moon, etc). Constraints included NASA's time frame for the moon landing ("before this decade [the 1960s] is out," from President Kennedy's famous speech) as well as more specific technical constraints such as vehicle size, weight, etc.

BLUE SKY

Create a vision with enough detail to be able to explain, present, and sell it to others.

The Blue Sky stage of the process involved Grumman developing their original proposal design for the lunar module. The proposal addressed the technical, contractual, and program management requirements of NASA's RFP, and featured a vehicle design comprising two parts: an ascent stage that serves as the crew compartment, houses the navigational and life-support systems, and is used to return to orbit after the landing; and the descent stage housing the landing gear and support systems, and that serves as the launch platform for the ascent stage. This initial design met NASA's requirements, but in the end very little beyond the basic two-stage design remained in the final design, and nearly every detail was changed later in the design process. For example, the proposal design featured a round descent stage with five legs and an ascent stage that included seats for the astronauts, large windows, and two docking hatches. The final version had an octagonal descent stage with four legs and an ascent stage with no seats, a handful of small windows, and only a single docking hatch. But while most (if not all) of the details changed during the subsequent design stages, the overall vision for Grumman's Lunar Module remained much the same throughout its design.

CONCEPT DEVELOPMENT

Develop and flesh-out your vision with enough additional detail to explain what needs to be designed and built.

The Concept Development stage took the form of what Grumman refers to as its "Preliminary Design" stage, where they developed and finalized the design based on mission specifications, requirements, and constraints. Hundreds of design changes were made during this stage, as the design team reworked and refined their initial proposal design. These design changes were based on specific mission constraints, the most significant of which was the overall weight of the CSM/LM. Weight was a major factor in most design considerations of both vehicles (it's a long way to the moon, and every ounce of weight shaved off the total allowed for more fuel and other consumables). Examples of these changes include the removal of the astronauts' seats and reduction in window size, changes to the docking hatches, and the use of lightweight shielding.

The initial proposal design for the LM included seats for the astronauts as well as large windows that would allow them to see where they were flying when landing on the Moon. The large cockpit windows each weighed several hundred pounds, but were needed in that design because the astronauts had to be able to see from their seats. At some point during this stage one of the engineers on the project questioned the need for seats and the LM designers realized that by removing the seats and having the astronauts stand, the astronauts were closer to the windows and could make do with much smaller windows.

Other design changes brought about by mission constraints had to do with the docking and access hatches on the lunar module. The original design called for two circular-shaped docking hatches on the ascent stage of the module. The first was on the top of the spacecraft and was used when the CSM docked with the LM during the trip to the moon. This hatch allowed access to the LM from the CSM. The second hatch was to be forward-facing, and would be used for egress to the lunar surface as well as docking with the CSM upon re-entry into lunar orbit. During preliminary design, the weight of the docking collars arose as a concern and caused a change to the design. The top-facing hatch remained as is, and would be used as the only docking connection to the CSM (this also necessitated adding a small window on the top of the LM to allow the astronauts to see the CSM when docking). The docking mechanisms were removed from the forward-facing hatch, which would be used solely for exiting/entering the LM from the lunar surface. The shape of this hatch also changed from round to rectangular to accommodate the shape of the Primary Life Support System (PLSS) backpack the astronauts would wear on the moon (another constraint). In addition, the way in which the door opened was determined by the mission requirement that the commander would be the first to exit the LM.

The last example we'll look at is the thermal and micro-meteorite shields. The types of thermal shields used on the Command Module were too heavy for the descent stage, so the

Grumman designers came up with a lightweight alternative. In a chapter called "Trimming Pounds and Ounces" in his book, Tom Kelly explains:

> Substantial weight was saved by redesign of the micro-meteorite shielding and thermal blankets that covered the exterior surfaces of the LM. We made a very efficient and lightweight combination consisting of fiberglass plastic standoffs bonded in the underlying structure, supporting a .005" aluminum alloy sheet of micro-meteorite shield, and insulation blankets consisting of multiple layers of aluminized mylar only one-eighth of a mil (.000125 inch) thick. The mylar blanket material was specially developed for this application.

Design

Develop the plans and documents that describe and explain how your vision will be brought to life.

The Design stage was Grumman's "Detailed Design" stage where engineering drawings and other design documents would be created. As Kelly explains:

> With the preliminary design completed, the focus in LM Engineering shifted to getting out the drawings, specifications, and other technical-definition documents that would enable Manufacturing to build the LM.

Early estimates for the number of drawings needed were in the thousands, but by the end of the project, the Grumman drafting teams had created more than 50,000 design documents. These technical drawings included renderings of the overall spacecraft as well as separate ascent and descent stages, but also had to document the design of the individual systems, sub-systems, components, and parts that made up the LM.

Another aspect of the Design stage was the design of various systems and tools used in the assembly and maintenance of the LM, known as "Ground-Support Equipment," or GSE. According to Kelly:

> Several hundred GSE items had been identified, and the list was continually growing. They came in a great variety, the most complex being the deliverable GSE end items. ... After use in LM final assembly...these end items were delivered to NASA with the LMs they supported and were installed and operated at Kennedy Space Center.

CONSTRUCTION

Build the actual project, based on the design developed in the previous stages.

During the Construction stage, the lunar module moved from Grumman's Engineering group to its manufacturing department, which was the team responsible for the physical construction of the vehicle.

A critical part of the construction of the lunar module was the fabrication of the parts from which the LM would be built. While the vehicle's design made use of some existing parts and components, such as fasteners, wires, cables, etc., most of the parts used in the LM had to be fabricated from scratch. Grumman's manufacturing team used different techniques and tools in the fabrication process, including traditional machining as well as chemical milling.

Once parts were created, the components, sub-systems, and systems of the LM could be assembled, including the vehicle's basic frame and bulkheads, interior and exterior components such as landing gear, propulsion and navigation systems, and life support, as well as its various electronics systems such as communications, radar, and avionics. Component parts came together into specific assemblies (interior cockpit panels and systems, window frames, hatchways), which in turn were assembled into larger pieces of the spacecraft, until the two stages of the LM were complete and could be fitted together.

Along the way, each component, assembly, and system was tested to ensure it functioned per mission requirements, and would perform reliably during actual missions in space. Grumman's teams performed thousands of tests spanning several years, revising and adjusting their designs as dictated by the test results. Extensive integration testing was conducted to ensure that the LM's various components and systems would work together as designed. Even though individual components and systems had been tested on their own, they also had to be tested when installed and integrated together in the spacecraft. For example, the radar system might have functioned as designed when tested in isolation, but it might perform differently (and unpredictably) once installed alongside the LM's communications and other electronics systems.

MODELS

Test and validate your design at each stage to help solve and/or prevent problems that may arise during the design and construction process. The Grumman design teams used different types of models at various stages during the development of the lunar module, including various design and presentation models, testing models and prototypes, and mockups.

As the design of the lunar module was refined during Preliminary Design (or Concept Development to us), design and presentation models were built to reflect the changes in the design for presentations to stakeholders from both Grumman and NASA. The first of these was based on Grumman's initial proposal design in 1962, featuring the round descent stage with five legs, and large forward-facing windows and circular docking hatch on the ascent stage. Later versions of this model created in 1963 and 1965 show changes to both the ascent and descent stages, including the now hexagonal descent stage with four legs, and smaller windows on the ascent stage. A final version of the model, reflecting the final design that would fly to the moon, was built in 1969. A picture of these four models side by side can be found on the "Lunar Module—External Design" page of NASA's Chariots for Apollo: A History of Manned Lunar Spacecraft" website (https://www.hq.nasa.gov/pao/History/SP-4205/ch6-2.html).

Grumman also built models and prototypes of various lunar module components and systems for use in testing. For instance, several prototypes of the LM's landing gear assembly were built to test various materials and designs for both the mechanism by which the landing gear would be extended prior to landing and for the landing itself.

In addition to design and testing models, the Grumman teams used three different mockups during their Preliminary Design and Detailed Design stages to uncover and resolve challenges in the spacecraft's design. According to Kelly:

> [The LM team concentrated] on three mockups during the first year of the LM program: M1, a wooden mockup of the ascent stage and crew compartment; TM-1, a wooden model of the complete LM; and M-5, a detailed metal model of the entire LM. ... We needed the mockups in order to fit check the LM's

complex shapes and assemblies and to determine whether astronauts and ground crew would be able to perform all required functions during the mission and during pre-launch preparation and maintenance.

EPILOGUE: OPENINGS, EVALUATIONS, AND SHOW QUALITY STANDARDS

Present your project to your audience, allow them to experience it, and evaluate its success and effectiveness over time.

The LM's "openings" were its initial test flights and "pre-moon landing" missions. These test flights and missions were done using completed lunar module vehicles. The first LM to fly was an unmanned test vehicle (LM-1) designed specifically to test the LM's performance in space. A second unmanned test flight was planned for LM-2, but the first was so successful that NASA dropped the mission and moved on to the first manned LM flight using LM-3. Code-named "Spider," LM-3 would be the first manned LM in space, and would fly on Apollo 9 with astronauts Jim McDivitt, Dave Scott, and Rusty Schweickart, while LM-4 would be used in a dress rehearsal of the moon landing on Apollo 10 when astronauts Tom Stafford and Gene Cernan flew to within 15 kilometers of the lunar surface. These missions provided valuable data used to refine the other lunar modules and later missions, including LM-5, the "Eagle" that would carry astronauts Neil Armstrong and Buzz Aldrin to their historic landing on the moon on July 20, 1969.

Imagineering Birthday Parties

The second example I want to look at in this chapter is a game I created for my son's 8th birthday party. This is a relatively small personal project, but I think it's another good example of the Imagineering Process at work.

For his 8th birthday, my son Nathan wanted a Star Wars-themed party. He was already a big Star Wars fan by then, and so in addition to the standard cake and ice cream, we planned for a number of Star Wars-themed activities, including light saber duels and a visit by Darth Vader. We had hoped that most of the party could be outside in our driveway and back yard, but as the date for the party got closer and closer, the weather forecast was looking grim. We soon realized that we needed to plan for indoor activities for the party, since having 8-year olds swinging toy light sabers inside our house wasn't going to fly.

I ended up creating a fun and simple Star Wars starship battle game called "Battle of the Death Star!" in which each child controlled a toy starship in a re-creation (of sorts) of the Battle of Yavin from *Star Wars Episode IV: A New Hope* (which will always just be "Star Wars" to me).

PROLOGUE: NEEDS, REQUIREMENTS, AND CONSTRAINTS

Define your overall objective, including what you can do, can't do, and must do when developing and building your project.

The Need was simple. We had to create a game that could occupy my son and the nine or ten guests we had coming to his party. The Requirements and Constraints for the project were based on the party's theme and attendees, and included things like:

- The game had to be based on Star Wars in some way.
- The game had be simple enough that I could explain it to 8- and 9-year old kids.
- The game had to be engaging enough to keep a group of nine or ten kids occupied for 45–60 minutes.
- We had to be able to put the game together in a few days.

BLUE SKY

Create a vision with enough detail to be able to explain, present, and sell it to others.

The Blue Sky stage for this project mostly involved my wife and I tossing around ideas such as an indoor scavenger hunt (we had done scavenger hunts at some of Nathan's previous parties that were well liked by our guests) or a card game of some sort. Nothing really seemed promising and then I remembered that my son had a set of Star Wars Micro Machines that I had bought years before when I worked as a game designer (and used them to decorate my desk and work area). When I bought the ships I had thought it would be cool to create a starship combat game using them, but had never gotten around to it. With the party closing in on us, now seemed like an ideal time to give that idea a try.

The idea was simple: I would create an easy-to-learn and simple-to-play Star Wars starship battle game in which the kids at the party would control space ships in a battle of some sort.

CONCEPT DEVELOPMENT

Develop and flesh-out your vision with enough additional detail to explain what needs to be designed and built.

Now that I had an idea, Concept Development involved me fleshing out what the game would be and how it would be played.

We had both Rebel Alliance and Galactic Empire ships, so it made sense that we would have two teams of players, but I didn't think a free-for-all battle would work well. We needed a scenario that would provide a goal for the players beyond "attack the other guys!" I realized that since we were talking about Star Wars, there was no better space better battle to re-create than the Death Star battle from the original Star Wars film. The Rebel Alliance players would be trying to destroy the Death Star before it can destroy the Rebel base while the Galactic Empire players would be trying to fight off the Rebels and let the Death Star crush the Rebellion forever.

In terms of game play, I wanted the game to be simple, but not simplistic. I didn't want a game like Candyland where there is no decision-making on the part of the player, but at the same time, it needed to be simple enough that I could explain it in a few minutes and the kids could understand how to play quickly and easily. I decided that each player would control a single ship, and that each turn a ship could move and fire its weapons at an enemy ship. That would provide opportunity for some basic decision-making (where to move and which ship to attack), but not be overwhelming. To help make sure the kids understood the rules, I would act as a referee, keeping things moving and reminding the kids how the rules worked along the way.

DESIGN

Develop the plans and documents that describe and explain how your vision will be brought to life.

The Design stage involved me working out the details of the game, including the basic rules as well as the specifics of each type of ship.

The rules of the game included how to decide which team would go first each turn, how the ships moved, how attacks would be resolved, what happened when a ship gets hit in battle, and how the Rebels could destroy the Death Star.

The specifics of the ships included how far each ship could move each turn, the range and accuracy of their weapons, the damage they inflicted if they hit an enemy with their weapons, how maneuverable they were (how hard they were to hit), and how much damage each could sustain before being destroyed. I spent a fair amount of time tweaking the ship statistics, since I wanted each ship to be different from the others in some way. For instance, one ship might move fast but not be especially durable, while others might not move fast but had more powerful weapons. I also had to balance the ships' statistics so that every ship was effective in the game. I didn't want any of the kids to get stuck with a "bad" ship. I even created "to hit" matrices that let me analyze the die rolls needed for each ship to successfully attack each type of enemy ship, just to make sure nothing was too out of whack.

I also designed the overall scenario of the game: The Death Star would start in one corner of the table (we played standing around our kitchen table) with the Rebel base in the opposite corner. Each turn, the Death Star would move 6 inches toward the Rebel base. After 10 turns, the Death Star would be in range to destroy the Rebel base. That meant that the Rebel team had up to 10 turns to destroy the Death Star. Once turn 11 happened, it would be Game Over for the Rebels.

CONSTRUCTION
Build the actual project, based on the design developed in the previous stages.

For the Construction stage, I wrote out the rules of the game (for my own reference during the party) and created record sheets for each ship that I could give to the kids. I wanted it to be easy for them to find the information they needed on the record sheets, and I gave them space to write their names and record any damage their ship took during the battle.

I also had to create or assemble other materials we would need to play the game, such as dice, pencils, and props for the

Death Star and the Rebel base. One specific item I created was a set of measuring strings that would be used during the game. Each ship could move a certain number of inches each turn, and each could only attack enemy ships within a certain number of inches. Rather than use rulers, I cut a set of 10-inch strings with markings at each inch. To measure movement and range, the player could put one end of the string at their ship and count a number of marks on the string equal to their movement rating or weapons range.

MODELS

Test and validate your design at each stage to help solve and/or prevent problems that may arise during the design and construction process.
In terms of Models, I created a couple of different versions of the record sheets before settling on the final version. I also playtested the game with my son and daughter before the party to make sure it would play well during the party.

EPILOGUE: OPENINGS, EVALUATIONS, AND SHOW QUALITY STANDARDS

Present your project to your audience, allow them to experience it, and evaluate its success and effectiveness over time.
As predicted, the weather was not great on the day of the party, so having this game ready worked out really well. Of course, given the work I had done on it, we were going to play it even if the weather was perfect!

We played the game in the middle of the party, after some opening games, but before the cake and ice cream. The kids all enjoyed the game. I even heard positive remarks from the parents of some of the kids who had told them about the game.

A few years later Nathan had another Star Wars party, and he decided he wanted to play the game again. I made a few minor tweaks to the rules and ship statistics to iron out some minor issues that came up during the first party, and the kids enjoyed the game this time, too.

CHAPTER THIRTEEN

Imagineering Game Design

In this chapter we're going to look at how the Imagineering Process can be applied to the design, development, and production of games. My own experience with game design is based on several years spent as a freelance game designer and in-house game designer and developer. My work was primarily focused on table-top roleplaying games, but I also worked on a handful of board games and card games. Because so much of my experience is based on roleplaying games, many of the examples in this chapter will be based on roleplaying game design, but I've included references to other types of games as well.

The chapter starts with a look at each of the stages of the process through the lens of game design, considers how the Imagineering Process compares with a traditional game design process model (the waterfall model), and provides an example of the process at work in developing a roleplaying game book.

Game Design and the Imagineering Process

This section explores how each stage of the Imagineering Process works when designing games. As we move through each stage of the process, I'm going to walk through an example of developing one of the books I worked on as the product line developer for the *Earthdawn* roleplaying game.

PROLOGUE: NEEDS, REQUIREMENTS, AND CONSTRAINTS

Define your overall objective, including what you can do, can't do, and must do when developing and building your project.

The Need for most game design projects starts very much like that of other projects—with someone identifying a desire or need to create a new game or game product. This desire or need can be based on anything that interests the game's creative team. For instance, the original Need behind *Irrational Game* (designed by author Dan Ariely and published by the Pressman Toy Company) might have started out as something as simple as "I want to create a card game based on psychological research case studies about rational and irrational behavior" while the original Need that led to the publication of the *Thunderbirds Co-operative Board Game* (designed by Matt Leacock and published by Modiphius Entertainment) might have started with "We want to design a cooperative board game based on the 1960s British television series *Thunderbirds*." In the realm of computer and video games, there is a similar Need driving every new expansion game or installment in a long-running series such as *Halo* or *Madden NFL* or *The Legend of Zelda*.

In the table-top roleplaying game field, it's common for publishers to produce supplemental materials for roleplaying games that expand on the game's world, provide new game rules, or provide pre-established stories or adventures for gamemasters to use in their games. Each of these supplemental products represents a Need, such as "We need to publish a book with new monsters for our fantasy roleplaying game" or "We need to publish an adventure for beginning characters for our new roleplaying game to show gamemasters what adventures in our game world are like." When I worked as a product line developer, often times these Needs started out quite generic, such as "We need a new sourcebook for the game."

Requirements for games are often based on its subject matter and how the game is intended to be played and experienced. For games based in a specific fictional setting such as the world of *Halo* or any number of table-top roleplaying games, accurately and engagingly portraying the game's world is a key requirement. For example, when I worked with authors on a book about dragons for *Earthdawn*, we weren't looking for generic information about dragons—the book had to deal with the specific types of dragons found in the *Earthdawn* world. In terms of game play, the style of game you're designing imposes certain kinds of requirements as well. For instance, a first-person shooter video game has different game play requirements than a real-time strategy game or a turn-based tactical game. In the area of table-top games, a competitive board game has different requirements than a cooperative game or a card game.

Constraints for games are often based on the type and format of the game, such as whether you're designing a board game, a card game, a computer-based video game, a console-based video game, a mobile game, a paper-based roleplaying game, etc. For example, card games have certain constraints based on the game's format (game play information has to fit onto a card), and a video game designed for mobile devices may have specific constraints (screen size, usability, etc.). For projects that are part of a larger series (whether that be a new installment in a video game franchise, a new expansion for a card game, or a new supplement for a roleplaying game), previous products in that series provide a different type of Constraint, since ideally you don't want to repeat what's been done in the past. Games also face constraints common to many types of projects, including time, money, and resources.

As you consider these aspects of your game, remember to make sure you've identified your real Need as well as your real Requirements and Constraints (see "Identifying Your Real Need, Requirements, and Constraints" in Chapter 4 for more about this).

To start my example of the *Earthdawn* dragon's book, at that time FASA was publishing four *Earthdawn* products per year, and when we began our planning for each year, we started with a simple and generic Need: "We need four new books for the *Earthdawn* game line."

BLUE SKY

Create a vision with enough detail to be able to explain, present, and sell it to others.

In this stage, you develop ideas that address your Need.

The specificity or vagueness of your original Need will play a big part in how you go about developing initial ideas for your game or game product. For instance, If your Need is generic ("We need a new sourcebook for our fantasy roleplaying game"), the lack of specificity provides for lots of possibilities: you could create a treasure book, a creature book, a book about cults and secret societies, a book about an important city, etc. In cases like this, a good approach might be to have full-fledged brainstorm sessions with no preconceived ideas of what you're going to produce and in which—like for the Disney Imagineers—the sky is the limit!

In some cases, however, the Need might be more specific, such as "We need to develop a new *Halo* game" or "We need to create a new *Magic: The Gathering* expansion." For projects like these, the specifics of the Need (and its Requirements and Constraints) may point you in a specific direction from the outset. At a minimum you would want to consider all of the previous products in the game's series to ensure you're not simply producing a clone of an existing product (unless, of course, that's the intent—but let's assume that's not the case here). A good question to ask is, "What aspect of the game (or the game's world) haven't we explored yet?"

As we noted in Chapter 5: Blue Sky, two important outcomes of the Blue Sky stage are defining your project's story (or subject matter) and its creative intent. Your story is what your game is about. Everything about your game should relate to or support its subject matter in some way. Knowing your game's story is critical. During each stage of the Imagineering Process, that story should be one of the primary sources of validation as you make decisions about the game. If an idea or element doesn't relate to or support the story in some way, it doesn't belong in your game. Specific Needs can help in determining your game's subject matter. For instance, if your Need is to "create a new game in the *Halo* franchise," that will dictate much about your subject matter.

As in the case of theme park attractions, a project's creative intent is what the designer wants to accomplish with the project and defines the experience the designer hopes to create for their audience. The primary objective of a game should be the experience that people who play the game will have. In *The Art of Game Design: A Book of Lenses*, author and former Imagineer Jesse Schell talks about the idea of experience:

> "What is the game designer's goal?" At first, the answer seems obvious: a game designer's goal is to design games.
>
> But this is wrong.
>
> Ultimately, a game designer does not care about games. Games are merely a means to an end. On their own, games are just artifacts—clumps of cardboard, or bags of bits. Games are worthless without people to play them. Why is this? What magic happens when games are played?
>
> When people play games, they have an experience. It is this experience that the designer cares about. Without the experience, the game is worthless.

Similar to its impact on story and subject matter, a specific Need can dictate much about the experience you want players of your game to have. For example, if your Need is something like "We need to publish an adventure for beginning characters for our new roleplaying game to show gamemasters what adventures in our game world are like," your creative intent will center on introducing players and gamemasters to your game and its world.

It's important that you know your story and creative intent before you move on to Concept Development and other stages in the process, as they both will help you refine your ideas as you progress into those later stages.

The ideas developed in the Blue Sky stage are often captured in a "Game Design Overview" document written so that stakeholders can know enough about what the game is, and who it is for, without getting into too much detail. The overview document can be useful for the whole team to get a sense of the big picture of the game.

In the example of the *Earthdawn* dragons book, we brainstormed different ideas, taking into account the books we had already published, and settled on a book about the dragons of the *Earthdawn* world. Our concept for the dragons book established both our subject matter and our creative intent. The book would be about the dragons of *Earthdawn*, providing general information about them, including the different types of dragons in the game's world, their powers and abilities, their servants and agents, how they interact with the other races in the game, etc. In addition, the book would also describe several of the named dragons in the game, and would provide *Earthdawn* gamemasters with the information they need to incorporate dragons into their games.

CONCEPT DEVELOPMENT

Develop and flesh-out your vision with enough additional detail to explain what needs to be designed and built.

So you have an idea for a game. That's great. But before you can design and create your game, you need to fully understand what exactly you want to create and what it will take to produce it. When you move into the Concept Development stage, you take your Blue Sky game idea and develop and expand on it so that you, and others working with you, understand what you have to design and build.

But where do you start when developing your game idea? Games are made up of many different elements, and knowing where to start can be difficult. One approach is to focus on what Jesse Schell refers to as the "four basic elements that comprise every game": Mechanics, Story, Aesthetics, and Technology.

- **MECHANICS** are "the rules and procedures of your game. Mechanics describe the goal of your game, how your players can and cannot try to achieve it, and what happens when they try."
- **STORY** is "the sequence of events that unfolds in your game. It may be linear and pre-scripted, or it may be branching and emergent."
- **AESTHETICS** is "how your game looks, sounds, smells, tastes, and feels. Aesthetics are an incredibly important aspect of game design since they have the most direct relationship to a player's experience."

- **TECHNOLOGY** refers to "any materials and interactions that make your game possible such as paper and pencil, plastic chits, or high-powered lasers. The technology you choose for your game enables it to do certain things and prohibits it from doing other things."

These elements don't stand apart from each other; decisions about each of these impacts the others. For instance, your choice of mechanics will inform aspects of the other three, since, as Schell explains:

> You will need to choose technology that supports [the mechanics], aesthetics that emphasize [the mechanics] clearly to players, and a story that allows your...game mechanics to make sense to the players.

Likewise Schell tells us:

> When you have a story you want to tell through your game, you have to choose mechanics that will both strengthen that story and let it emerge. ... You will want to choose aesthetics that help reinforce the ideas of your story, and technology that is best-suited to the particular story that will come out of your game.

Your original Need, Requirements, and Constraints will play a large role in determining how you develop these elements. For instance, if you know you're designing a card game, your technology is defined and will impact your mechanics, aesthetics, and story. If you're developing a supplement for a roleplaying game, your technology and aesthetics will be pre-determined, but you may have more leeway in terms of mechanics (such as new rules or new applications of existing rules) and story. In addition, some of these elements may have multiple facets to them. For instance, your story might involve developing characters and settings, while your mechanics may involve multiple sub-systems or exceptions to the general rules.

For many games and game projects, the output of the Concept Development stage is a document that outlines the details of the project in sufficient detail so that other designers can understand the game, its subject matter and objectives. As Schell describes it:

> [The "Detailed Design Document"] describes all of the game mechanics and interfaces in great detail. This document usually serves two purposes: so the designers remember all

the little detailed ideas they came up with, and to help communicate those ideas to the engineers who have to code them and the artists who have to make them look nice.

Another name for this document is simply a "game design document" which should describe the game's selling points, target audience, gameplay, art, level design, story, characters, UI, assets, etc. In the case of roleplaying game design, the document might be a detailed outline that describes the project with sufficient detail for designers and writers to write and design the game, sourcebook, or adventure.

In the example of the *Earthdawn* dragons book, at this stage I developed a detailed outline of what we wanted the book to contain, including a list of the specific named dragons we wanted to highlight and the specific information about each we wanted to include, an overview of the general information about dragons we needed to cover, notes about the game rules we needed to provide and/or elaborate on, etc. This required me reviewing our existing *Earthdawn* products to note what we had previously published about dragons, including game rules, stories that featured one or more of our named dragons, and other bits of history and lore.

I also worked out what the book would be within the game's world. One of the conceits of the *Earthdawn* product line was that many of the books we published were from the *Earthdawn* game world. For example, several published books were said to be from the "Library of Throal," the library of the large dwarf kingdom that was effectively the capital city of the game's main setting. In the case of the dragons book, it would comprise two parts. The first part would be an essay about dragons written by one of the Great Dragons (transcribed by a human servant) that was to be donated to the Library of Throal. The second part would be a series of essays about several specific named dragons written by one of their own who had gone rogue and had violated several rules of dragon society. These two different points of view allowed us a rationale to provide interesting and "secret" information about the dragons of the *Earthdawn* world.

Once I completed the outline, I provided it to several authors who submitted proposals to be part of the writing team. Based on the proposals I received, I assigned a number of authors to specific sections of the book. Most author assignments were descriptions of one or more of the "named" dragons that comprised the second half of the book, while one lead author also wrote the section that provided general information about dragons in the *Earthdawn* game world.

DESIGN

Develop the plans and documents that describe and explain how your vision will be brought to life.

The Design stage is where you start to design and build your game. This is where designers and contributors from different specialized disciplines get involved in the design process. For example, designing video games involves designers from several disciplines such as level design, technical design, system design, user interface (UI) design, tools design, sound design, etc.

Table-top games, including board games, card games, and roleplaying games, involve not only designers working on the game rules and text, but also graphic designers and illustrators who can help bring the game's aesthetics to life. As in the case with Imagineering and theme park design, designers from varied disciplines have to collaborate to make sure their work fits together to bring the game to reality. For example, in the case of card games and board games, the people designing the rules and mechanics must work with graphic and visual designers to make sure that the components are designed so that the information they contain is communicated and conveyed effectively.

Game design is a field in which the Design stage often merges with the Construction stage as work on projects proceeds through the process. For instance, some of the design work on a video game might also include coding and engineering of prototypes and proofs of concepts that end up being incorporated into the final game.

In the case of roleplaying games, this stage is where writers and designers write and develop the text, mechanics, and game statistics that will form the book's content, whether it be a rulebook, a sourcebook, or an adventure. In addition, the art team might start soliciting sketches and ideas for the book's cover artwork. The same detailed outline that is provided to writers and designers is typically provided to an art director, who in turn provides it to the cover artist so he or she can better understand what the book will be about.

For the *Earthdawn* dragons book, the Design stage is where the individual authors wrote and designed their portions of the book. They submitted first drafts to me for review, I provided feedback and notes for revisions and updates, and then they submitted the final drafts that I would assemble, edit, and develop into the final book.

CONSTRUCTION

Build the actual project, based on the design developed in the previous stages.

The Construction stage is where the various pieces, parts, and components of your game are developed and assembled into the final product. As noted above, some of the work typically thought to be part of this stage may start in the Design stage and continue through to Construction. As is the case in the Design stage, the specific types of work done here will vary based on the type of game you are creating.

When developing video games, this stage is where most of the software coding happens, including implementing mechanics (how the rules manifest in game play), 3-D modeling (how objects are rendered in the game), object interaction (how objects in the game interact), etc. In addition, visual artists create backgrounds and art work, and sound engineers record and create the game's sounds, sound effects, and recorded dialog. This is obviously a simple and high-level overview of what goes into the construction of a video game, but is (hopefully) sufficient for our purposes.

For most table-top games (including board, card, and role-playing), the game's mechanics and rules are often designed and written during the Design stage. The Construction stage is where those elements are brought together. For example, the Construction stage for board games involves the integration of the game's rules, mechanics, and statistics with its physical components—the board, cards, tokens, etc. Construction of card games is similar, but limited to incorporating the game rules, mechanics, and statistics on different types of cards, such as creature cards, resource cards, etc. When creating roleplaying products, this stage picks up after the content has been designed and written, and involves story and rules editing, editing for style and language, book design and layout, indexing, printing, and assembly.

The "Test and Adjust" part of this stage primarily involves playtesting. Schell defines playesting as "getting people to play your game to see if it engenders the experience for which it was designed." The specific manner in which you playtest will vary depending on the type of game (for instance, you would playtest a board game differently than you would a card game or roleplaying game or video game), but every playtest should have a specific objective of its own, and should be defined by the five questions that form Schell's "Lens of Playtesting":

> Playtesting is your chance to see your game in action. To ensure your playtests are as good as they can be, ask yourself these questions:
>
> - **Why** are we doing a playtest?
> - **Who** should be there?
> - **Where** should we hold it?
> - **What** will we look for?
> - **How** will we get the information we need?

The Construction stage of the *Earthdawn* dragons book started with me assembling the parts of the book submitted by the different writers, and doing what we called a "Development" edit where I made sure the details of the *Earthdawn* setting, as well as the new rules and mechanics in the book, were correct and accurate. Once I was done, the book moved on to the editorial department where the text was reviewed and edited for our house style, proper English and grammar, etc. The book then moved to the art and production department, where it was laid out, art work was commissioned, and the text underwent a final proofread. (Unfortunately, this book was never actually published, and work on it stopped at this stage before the art work was received and inserted into the final layout.)

MODELS

Test and validate your design at each stage to help solve and/or prevent problems that may arise during the design and construction process.
As described above, one of the primary types of models used in game design is playtesting. Games can go through several rounds of playtesting, starting with testing of individual components before moving on to testing of the "final" product (also known as "beta" testing). For instance, when designing a roleplaying game, you might playtest the combat system at a fairly early stage in the game's development to make sure that basic elements of the system work as intended.

In addition to playtesting, game design also often involves the use of prototypes and mockups. When designing card games, card mockups range from cut out slips of paper, to slips of paper taped to playing cards, to cards printed on blank business card stock. When working on board games, prototypes and mockups can include hand-drawn paper maps, stand-in components (plastic poker chips or bingo markers), card mockups, etc.

In the design of roleplaying games, models can take the form of outlines, storyboards (used to visualize the flow of events in an adventure, for instance), and drafts of the game text and rules.

The Models we used in the development of the *Earthdawn* dragons book included first drafts of the individual author contributions, sketches and ideas for the book's cover, and the initial rough layout of the book.

EPILOGUE: OPENINGS, EVALUATIONS, AND SHOW QUALITY STANDARDS

Present your project to your audience, allow them to experience it, and evaluate its success and effectiveness over time.
Openings for games are when the game is published and released. In the case of video games, games are released via game stores (both brick & mortar and online) or via online services such as Steam. Board games, card games, and roleplaying games are usually released in stores, but in recent years, many roleplaying games are initially released in digital format, followed by release in hardcover or softcover sometime later.

Some games launched via crowdfunding sites like Kickstarter provide previews for backers in which those who backed the game early in its development have an opportunity to preview the game before it is available to the general public. In addition to giving backers early access, in some cases these previews also provide an opportunity for the game's designers to update and revise the game based on feedback.

Once released and in the hands of players (and gamemasters for roleplaying games), evaluations happen quickly. Game fans are generally not shy about sharing their opinions about games and game products, including posting online reviews on websites and blogs, creating unboxing and actual play videos, and discussing the game's merits and flaws (few games are flawless) on online forums. In addition to taking in public feedback, many game publishers perform their own evaluations of their products with an eye toward improving the game and the players' experience.

Roleplaying game publishers often publish errata that address any errors or issues (both minor and major) that made it into the published version of the game. For video games, bug fixes and patches are often made available through the publisher's website or other online services (such as the Apple and Google app stores in the case of mobile games).

Successful games may have subsequent editions published which feature major or minor updates and changes to the game based on player feedback. This is common in the roleplaying game industry. For example, *Dungeons & Dragons*, the most popular roleplaying game in the world, is currently in its 5th edition; *Call of Cthulhu*, a popular horror roleplaying game based on the writings of H.P. Lovecraft, is currently in its 7th edition; and *Earthdawn* is currently in its 4th edition. Note that while these follow-up editions are born from feedback and evaluations, they also each represent a new Need, bringing the game's designers full circle from the Epilogue of one project to the Prologue of another.

As noted above, the *Earthdawn* dragons book was not formally published by FASA. The company ceased production of the game a few months before the artwork for the book was completed. A few months later FASA released a PDF version of the book based on its initial layout (this version contained no artwork). A company called Living Room Games licensed the *Earthdawn* game from FASA and later published a revised version of the book as part of their *Earthdawn Second Edition* product line.

Imagineering Waterfalls

Another way to look at applying the Imagineering Process to game design is to compare and contrast it to other game design process models. One specific process model commonly used in the video game industry comes from the traditional software development world and is known as the the "waterfall" model.

Jesse Schell describes the waterfall method as an "orderly seven-step process for software development...generally presented looking something like this":

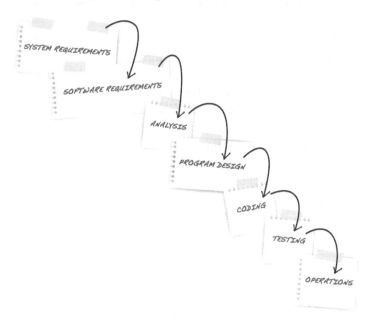

Schell writes:

> It certainly looks appealing. Seven orderly steps, and when each is complete, nothing remains but to move on to the next one—the very name "waterfall" implies that no iteration is needed, since waterfalls do not flow uphill.

However, looks can be deceiving, and Schell is not shy in sharing his views about the issues associated with the waterfall method:

> The waterfall method had one good quality: it encouraged developers to spend more time in planning and design before just jumping into code. Except for that it's complete nonsense...software is simply too complex for such a linear process to ever work.

Schell is not alone in his criticism. The description of "waterfall development" on the *What Games Are* website notes:

> [Waterfall] is a widely hated process of working, especially in games, because it treats the project as a set of certainties and development as essentially an act of translation. The reality of most game projects is that they are often trying to discover a game dynamic which works (called "finding the fun"), which is an inherently sloppy and experimental process.

However, despite the challenges inherent in the waterfall method, it remains a popular approach to video game development. According to *What Games Are*:

> Most contracts between developers and publishers are formulated in a quasi-waterfall fashion, mandating milestones such as game design documents and deliverables such as assets or functionality on a timetable. The reason they do it this way is that waterfall tends to give the illusion of certainty and is the easiest method to explain at the executive level.

So, how does the waterfall method align with the Imagineering Process? A comparison between the two looks something like this:

- The waterfall System Requirements and Software Requirements stages are roughly equivalent to the Imagineering Process Prologue and its Needs, Requirements, and Constraints.
- The waterfall Analysis stage is parallel to the Imagineering Process Blue Sky stage, where we develop an idea that addresses the requirements.
- The waterfall Program Design stage is a combination of the Imagineering Process Concept Development and Design stages, where the initial idea is further developed and fleshed out, and design documents are created.

- The waterfall Coding and Testing stages together are the Imagineering Process Construction stage, where the project is built and tested.
- The waterfall Operations stage is the Imagineering Process Epilogue and our Openings, Evaluations, and Show Quality Standards.

I believe adopting some of the ideas and principles of the Imagineering Process can make the waterfall method more effective. Let's look at a couple of ways we can do this.

First, focus on the objectives of each stage of the Imagineering Process as you progress through the corresponding waterfall stages. The following table outlines how these objectives align with the waterfall stages:

When working on:	You should:
System and Software Requirements	Define your overall objective, including what you can do, can't do, and must do when developing and building your project.
Analysis	Create a vision with enough detail to be able to explain, present, and sell it to others
Program Design	Develop and flesh out your vision with enough additional detail to explain what needs to be designed and built, and develop the plans and documents that describe and explain how your vision will be brought to life
Coding and Testing	Build the actual project based on the design developed in the previous stages
Operations	Present your project to your audience, allow them to experience it, and evaluate its success and effectiveness over time

Second, use the Imagineering Process checklist questions as you move through the stages of the waterfall process to maintain and refine your focus on these objectives.

Lastly, look for ways to incorporate the Imagineering Process' flexibility, scalability, and adaptability (see Chapter 11) to your waterfall process. For example, one thing that's noticeably missing in the waterfall method is iteration. Iteration is important in any creative process, including game design. As Schell explains, "The process of game design and development is necessarily iterative, or looping." He summarizes this idea with what he calls the Rule of the Loop: "the more times you test and improve your design, the better your game will be."

One More Roleplaying Example

To finish our look at applying the Imagineering Process to game design, I want to explore one more example from my time working on *Earthdawn*.

One of the more unique aspects of *Earthdawn* is its treatment of magic, and in specific, its treatment of magical items. In many roleplaying games, if a character finds a magical item, they can often pick it up and start using its magic right away. In *Earthdawn*, before a character can use a magical item, he or she must first learn about the name and history of the item and then create a magical connection between themselves and the item (via magical "threads"). Researching and learning the history of magic items can provide inspiration for all sorts of stories and adventures, and we wanted to publish a supplement that was an example of how to incorporate magical items into a campaign. The product we ended up publishing was an adventure collection called *Blades*, which includes a set of adventures in which the player characters research and discover the history of a set of cursed magical daggers, the Blades of Cara Fahd.

Let's look at how this product was conceived, designed, and developed through the stages of the Imagineering Process.

I already described the **Need** that drove this product: "We need to publish a supplement that provides an example of how to incorporate magical items into a campaign." The product's **Requirements** flowed from this basic Need and the nature of how magic items work in *Earthdawn*. The product had to provide one or more adventures which would allow the player characters to learn important pieces of history about a magical item. The product also had to provide gamemasters with game information and the history of the magical item. The project's **Constraints** were primarily related to the production costs and schedule for the book—we had a limited budget for writing and art work, and we had to complete the product by a predefined release date.

During the **Blue Sky** stage, I decided the product should be an adventure collection that gamemasters could use as the player characters sought to learn more about the history of the magic item. The book would include an adventure for each important piece of history the player characters would need to uncover in order to make use of the magic item. Rather than run these adventures one after another, we would recommend that gamemasters intersperse them with others in their campaign.

During the **Concept Development** stage, I started with designing and developing the magic item that would be featured in the adventure collection: the Blades of Cara Fahd, a set of matching (and cursed) magical daggers that had played a key role in the history of the ancient ork nation of Cara Fahd. The choice to use a set of magical daggers came from a desire to provide a magical item that could be of use to multiple characters, unlike a single magical sword or suit of armor that would only benefit a single character. We wanted the blades to be a part of the *Earthdawn* world and its ongoing story (its "metaplot"), so we first introduced them in an adventure called *Shattered Pattern*, published several months before *Blades*. This provided a rationale for how the player characters might have come to possess the blades.

I also drafted a detailed outline of the overall book, including an overview of the blades' history, their curse and the game impact the curse would have, and a brief description of the specific information the player characters would learn during each adventure.

I sent the outline to prospective authors, and based on the proposals I received, selected the five authors who would write the individual adventures.

During the **Design** stage, the authors designed and wrote their adventures, checking in with me as needed. Once they completed their first drafts, they submitted them for review and feedback. I also worked with the art department as they selected a cover artist and started to work with the artist on sketches and ideas for the book's cover.

The **Construction** stage started with the submission of the final drafts of the individual adventures. I read through each adventure again, making edits and changes to each based on the game's setting and mechanics, but also to ensure the adventures worked not only as individual adventures, but also as parts of a larger story.

As part of this effort, I identified a new Need and Requirement that I hadn't considered when we started the project. In addition to the five main adventures, the book also had to include a set of "interludes" between the adventures that served to link them together and provide the characters the information they needed to continue their research into the history of the blades. As I edited the individual adventures, I also wrote these interludes, as well as the book's introductory text.

The Construction stage also included the book going through FASA's editorial department where it was edited for style, grammar, etc., and FASA's art and production department where the book was laid out, its art work was commissioned and inserted, and it was ultimately sent to the printer.

The **Models** we used with this project included the proposals submitted by the individual authors, playtesting the authors did on their individual adventures, and their first and final drafts.

For the book's **Opening**, it was released and sold through hobby and game stores. It was generally well received by the *Earthdawn* fan community, but it was never revised or updated.

Imagineering Process Checklist Questions—Game Design

Stage	Questions
Prologue: Needs, Requirements, and Constraints	<ul><li>What is your game about?</li><li>What type of game do you want or need to design?</li><li>What are the game play requirements?</li><li>What type of constraints do you have to work within?</li></ul>
Blue Sky	<ul><li>What is your game's story?</li><li>What is the experience you want players of your game to have?</li><li>Have you developed your vision enough to be able to explain it to others?</li></ul>
Concept Development	<ul><li>Have you defined Schell's basic elements of your game: Mechanics, Story, Aesthetics, and Technology?</li><li>Have you developed your game concept enough that you could turn it over to someone else and they could take the next steps?</li><li>Is your concept feasible?</li></ul>
Design	<ul><li>What are the disciplines involved in designing your game?</li><li>What types of design documents do you need to build your game?</li><li>What specific types of design do you need to include in order to be able to implement your game?</li><li>Has your design work uncovered new Needs, Requirements, or Constraints that you hadn't identified previously?</li><li>Are there macro and micro levels of design for your game?</li></ul>
Construction	<ul><li>What are the tasks involved in assembling and constructing your game?</li><li>What sort of "Test and Adjust" effort are you using with your game?</li></ul>

Models	<ul><li>What types of models can you use as you design and build your game?</li><li>Are you playtesting your game? Are you playtesting it enough?</li><li>Are you creating prototypes and mockups of your game?</li></ul>
Epilogue: Openings, Evaluations, and Show Quality Standards	<ul><li>Are you able to offer a preview release of your game to a select audience (such as crowdfunding backers)?</li><li>How will you handle errors and issues that are uncovered after the game has been released?</li></ul>

CHAPTER FOURTEEN

Imagineering Instructional Design

In this chapter we're going to look at how the Imagineering Process can be applied to instructional design. For our purposes, any time you design or create experiences with a goal of teaching something, you're engaging in instructional design. This can take different forms, such as the development of corporate training materials or presentations, or designing lessons for school classrooms.

As I mentioned in the Pre-Show, the concepts in this book first took the form of a presentation I gave at a training conference about applying Disney theme park design principles to instructional design. While the connection between instructional design and Imagineering may seem odd at first glance (I can learn about teaching at Disneyland?), I believe the two are very well aligned because communicating ideas to an audience is at the heart of both fields. In the case of Imagineering, it's about communicating a story to park guests, while in instructional design it's about communicating new skills and knowledge to students.

In this chapter we'll look at each of the stages of the Imagineering process from an instructional design perspective, and then look at how the Imagineering Process compares and aligns with a traditional instructional design model known as ADDIE.

Instructional Design and the Imagineering Process

This section looks at each stage of the Imagineering Process with a specific focus on how the stage applies when designing learning and instructional materials.

PROLOGUE: NEEDS, REQUIREMENTS, AND CONSTRAINTS

Define your overall objective, including what you can do, can't do, and must do when developing and building your project.

Needs for instructional design projects begin when someone identifies and articulates the need to teach someone something. For example:

- We need video-based training courses for our consulting group.
- We need implementer training for our new software application.
- We need a comprehensive on-boarding training program for new hires.
- We need to expand an existing class to meet our state's revised curriculum standards.
- We need new technology and engineering classes to supplement our STEM/STEAM curriculum.
- We need new professional development sessions for our faculty.

Notice that all of these examples outline either teaching a specific subject to a specific audience or delivering instruction in a specific manner to a specific audience. These specifics, whether they be format, subject matter, or audience, form the core objective and basic Requirements for the project, but we must know more than the basics before we can start to identify a way to meet our Need.

Fully understanding the Need and Requirements of an instruction design project is the reason that a common initial step in most instructional design processes is known as a needs analysis. Common questions asked as part of a needs analysis include:

- What objectives/goals/problems should the training address?
- What should students know or be able to do after taking this course?
- Who is the target audience?
- What are the topics that must be covered?

Answers to these (and possibly other) questions can help you identify the Requirements for your project.

Like all creative projects, in addition to Requirements, many instructional design projects have Constraints. Common instructional design Constraints include the time, money, and resources necessary to develop the project ("We need the course completed by next quarter"), but other Constraints might include the duration of the course or lesson ("This new training course can be no longer than two days"), or the manner in which the course is delivered ("This training must not be lecture only").

Notice that in some cases, Constraints and Requirements can be thought of as two sides of the same issue. For example, a "not lecture only" Constraint might also be stated as "must include both lecture and hands-on practices"—a Requirement. Likewise, if one of the Requirements for a particular class or training course is that it must be stand-alone and not rely on any other pre-existing material, that Requirement ("the course must be stand-alone") is also a Constraint ("the course must not rely on other training").

BLUE SKY

Create a vision with enough detail to be able to explain, present, and sell it to others.

In the Blue Sky stage, you develop ideas that address your Need. At a minimum, this involves identifying what your audience needs to know, and the means by which you plan to teach them. Some of this may be dictated by your Need, while some may be undefined as you start your Blue Sky work.

As we noted in the previous chapter, the specificity or vagueness of your original Need can play a big part in how you go about developing initial ideas for your project. In particular, the specifics and details of your Need, Requirements, and Constraints can impact the amount of freedom you have at this stage. In some cases, by the time you start your Blue Sky stage you may be somewhat limited in terms of how you can address your Need, Requirements, and Constraints. For instance, a Need such as "We need to develop video-based training on basic first aid for our factory employees" leaves only so much room for creative solutions to the challenge. This is not to say that meeting this Need requires no creativity. It's just that very specific and detailed Needs such as this example can dictate a lot about the initial vision and concept for your project. In cases like this, the Concept Development and Design stages might offer more places for creative approaches.

On the other hand, a generic or vague Need leaves the door wide open for all sorts of creative ideas. For example, a Need like "We need new professional development sessions for our faculty" can be met by any number of different approaches, including both traditional instructional design methods such as instructor-led training, video-based lessons, e-learning modules, etc., and some not-so-traditional methods such as an interactive skit in which improv actors work from a skeleton script and solicit input from the audience at certain points—a sort of "choose your own training" session.

Two important aspects of defining what your audience needs to know are your project's story (or subject matter) and its creative intent. In instructional design, your story is the primary subject matter around which your learning experience is designed. Are you teaching about chemical reactions, the history of American theater, or how to install software on a computer? Your project's subject matter should be one of the primary sources of validation as you make decisions during later stages of the Imagineering Process. If an idea or element doesn't relate to or support your subject matter, it doesn't belong.

Most instructional design projects start with a Need based on a specific subject or topic, so part of your concept design work during this stage might be to refine how you define your project's subject matter so it's not too narrow but also not too broad. Too narrow and the usefulness of your course or lesson may be limited; too broad and you run the risk of trying to cover too much material.

In instructional design, your creative intent is the specific educational goal of the course—are you teaching functional knowledge, technical knowledge, etc.? Are you teaching someone how to answer customer support calls? Are you teaching chemistry or physics or history? Are you teaching somebody first aid? Are you teaching them how to implement a product? Your target audience should play a major role in determining the objective of your course. For example, an end-user training course has a different objective than an installation or implementation course. Everything you do should add something significant to the learning experience and should serve your creative intent.

Another important element to consider at this stage is the means by which you plan to teach your audience what they need to know, or put more simply, the format of your course or lesson. As we noted earlier, most instructional design projects are based on a specific subject or topic, but the method by which the subject matter is taught is not always specified. Should your project be instructor-led training or lessons, self-paced online learning, video-based lessons, e-learning delivered through a Learning Management System, or some other format? If your project is going to be part of a larger library of content, do you have a standard format that you should adopt for this project? If your Needs, Requirements, and Constraints don't specify anything about the format of your course, this is an area where you'll have more freedom during this stage.

As you consider your format, remember that there are many ways of teaching. When we think of instructional design or training, we often default to traditional training classes, classroom instruction, video-based lessons, and other more standard approaches. Don't limit your thinking by focusing on only these typical types of training and instruction. In recent years there have been a handful of new approaches worth considering. One of these is "gamification," or using "game thinking and game mechanics in non-game contexts to engage users in solving problems and increase users' contributions." Using gamification in instructional design has become quite popular recently as more and more designers look for ways to involve and engage their audiences. Adding game-like activities to your educational experience is a great way to keep the experience more dynamic and active. Another approach gaining popularity in recent years is "microlearning" or "short bursts of focused 'right-sized' content to help people achieve a specific outcome." When evaluating formats for your project, remember your objective: at a minimum, you need to identify what your audience needs to know, and the means by which you plan to teach them. What is the best way to convey to your audience what they need to know? In many cases, it might well be a traditional instructional method, but not all instructional design challenges require a class or lesson. Some might be better met by a job aid, or a checklist, or a game, or an infographic.

In the end, this stage should result in enough detail so that your stakeholders (including whoever it was who presented you with your Need) understand what you plan to build, and how you plan to address your Need. This could be as simple as a sentence or two, or as elaborate as dozens of illustrations, text descriptions, and storyboards presented in a multi-media presentation. Once you can effectively communicate your vision for what you plan to create to yourself and others, it's time to move on to the next stage.

CONCEPT DEVELOPMENT
Develop and flesh-out your vision with enough additional detail to explain what needs to be designed and built.
In Chapter 6 we noted that the Concept Development stage is where the initial concept creating during the Blue Sky stage is further developed and fleshed out in enough detail so that more detailed design work can begin. Even if you have a clear concept and vision for your instructional design project, there are probably lots of details and specifics that you need to work out before you can effectively begin to build your project.

As you begin this stage, consider your Blue Sky concept and look for aspects that are vague or undefined. For example, if you're designing onboarding training for your company, during Blue Sky you might summarize your concept as "develop training content in a variety of formats that provide new employees with the information they need to be effective in their jobs." That description is vague, and doesn't provide enough detail for you to start designing the training; you need to know a little more about the courses you plan to build first. In Chapter 6 we looked at some example questions that might apply here, such as:

- What specific topics should each course or module cover?
- Are there specific audiences for each course or module?
- What systems and tools do employees use in their work?

As you answer questions like these, you'll better understand the scope of the subject matter you need to address.

This stage is where you'll likely need to do a fair amount of research. Unless you're a subject matter expert (SME) yourself, you'll want to find one or more SMEs who can help you learn more about your subject matter. For example, if part of your onboarding training involves teaching employees how to use a variety of self-service HR tools, you would want to meet with SMEs from the HR department to identify the core functions these tools offer.

Your Concept Development work should also include iden-tifying the number of courses you need to develop. Some projects might require no more than a single two-hour course, while others might end up being several large courses, each of which comprises dozens of lessons and hundreds of topics. One approach to this is to organize your subject matter by similar topics. As you gather data about your subject matter, capture the specific topics you need to address and organize them into cat-egories (in the case of onboarding, they might include company policies, HR tools, and IT resources). From these categories you can further organize your topics into sets of individual lessons, modules, and courses. A simple method for this is to write the name of each topic on an index card, arrange the cards into piles sorted by your categories, and then further refine each category pile by creating smaller piles for each course, module, or lesson.

Once you have identified the set of courses you need to develop (along with some ideas regarding the specific lessons or modules in each), you should create high-level outlines and descriptions for each. These outlines can help you further organize your content. Storyboards and mind maps are great tools to use at this stage, as they can help you visualize how the content areas of your course are organized. They can also help you work out the best or most effective order in which to organize your content. In some cases the order in which you present information is based on basic prerequisites. A fundamental aspect of designing learning or instructional materials is identifying fundamental vs. advanced subject matter. For example, when teaching grammar, you wouldn't teach compound sentences before teaching simple sentences and the basic elements of grammar. Planning the order in which you will present your topics is related to the principle of Transitions from the Imagineering Pyramid. In the realm of theme park design, Transitions are about making changes as smooth and seamless as possible. In instructional design, Transitions apply to the order and sequence in which we want to present our topics, and identifying the "best" order or sequence.

When creating mind maps or storyboards, consider your content from both a macro and micro point of view. At the macro level you might create a mind map or storyboard for the overall project. At the micro level you can create individual maps for each of the individual courses, modules, and lessons.

In addition to outlining your subject matter, these outlines and descriptions should also indicate the best ways to convey the information in each course. You identified the training format (or formats) you want to use for your project during the Blue Sky stage, but here you want to get more specific and decide on the specific instructional method you want to use for each course, module, or lesson. For instance, teaching new employees about company policies is perhaps best done via video-based lessons, e-learning modules, or microlearning, while teaching them tasks such as how to create a timecard is probably best handled through a hands-on exercise or a simulation.

Your work in this stage should result in a set of one or more high-level outlines (or storyboards or mind maps) that describe your project (and its individual parts) in enough detail so that you and your stakeholders fully understand what needs to be designed and built in the next stages.

DESIGN

Develop the plans and documents that describe and explain how your vision will be brought to life.

In the Design stage you use the high-level outlines, mind maps, or storyboards created in the Concept Development stage to develop detailed design documents that clearly define and describe how your project will be built.

The documents you develop at this stage will be used by whoever will be building your project. In some cases that might be you or a team working with you, but in some cases you might be handing off these documents to other teams whose job it is to create and develop the instructional materials. Regardless of who will be creating the materials, you need to provide them the details they need to be able to do so effectively and efficiently.

In previous stages you had to rely on subject matter experts to help you develop your ideas and make sure they address your original Need, Requirements, and Constraints. This stage is where you get to put your instructional design expertise and knowledge to work. For instance, when creating a course about accounting software you may not understand all of the details of accounting, but when you get to this stage you can apply your experience to translate accounting-specific topics and subject matter into an effective instructional experience.

One type of design document you might create at this stage is a detailed outline that breaks down each part of your project into the individual pieces you need to create. You might start with a break down of the individual courses (if more than one). Each course might comprise a number of modules, each of which might comprise one or more lessons, which in turn might each include multiple topics, each of which can be broken down into a set of concepts. The size and scope of your project will dictate how extensive this outline will be. For simple projects, the outlines you created during Concept Development might be detailed enough, but for more complex projects, a detailed design outline could be three to four times the size of a concept outline. For example, if your materials will use presentation slides (such as PowerPoint), your detailed outlines might be so detailed as to list every slide you need to create. These outlines can also indicate different instructional methods for different parts of your project, such as lecture, hands-on practice, quizzes, etc. These outlines need not be traditional written outlines. Using mind mapping for detailed outlines can help you to step back and see the entire course and how its various parts flow together.

Another type of design document you can use during this stage is annotated storyboards that feature illustrations or diagrams of key topics, along with detailed notes providing specifics. Storyboards of this type are especially useful when designing video-based content, particularly if the video will feature actors or other scripted content. For example, when developing a first aid training video or a video on the use of fire extinguishers, your storyboards might be very similar to those used by filmmakers and Disney animators. When designing instructor-led training, storyboards can be used to map out an entire classroom experience (including lectures, quizzes, exercises, etc.), and also as a visual tool to see the entire course (similar to a mind map). For example, you might use three different colors of index cards for different types of content (white = lecture, green = quiz, blue = exercise), with each card containing a title or phrase describing each piece of the course. By putting those on a board and looking at it from a distance, you can see if you have too much of one type of content vs. another ("we have too much lecture here, we should spread this out," etc.).

Another type of design document you might need are scripts that dictate how specific tasks or procedures are performed. For instance, if you plan to include hands-on practices or exercises, you need to design those practices and write scripts explaining how to perform the practice or exercise, including specific actions to take at each step. So, if you want to include a hands-on practice around creating a time card using an HR application, you might outline each step of the task as follows:

1. Select Menu > Time Card > New.
2. Select the date range for the time card from the Date Range drop-down list.
3. Select the type of time entry from the Time Entry Category drop-down list.

4. Enter the appropriate number of hours for the category in the fields shown for each date in the selected date range.

In addition, if your exercises require specific data or materials, your design documents need to define them. In our time card example, you might specify the time entry categories employees will use when creating time cards (Regular Work Hours, Vacation, Sick Time, etc.).

If you plan to include recorded content or demonstrations, in addition to scripting out the steps to be performed (as above), you may also need to write a script for any narration that will accompany the demonstration.

As you work on your detailed outlines and other design documents, you may also wish to further refine the order and sequence of your content and examine the Transitions within your courses, modules, and lessons. We talked about Transitions under Concept Development when we noted that in some cases the order in which you present information is based on basic prerequisites (simple sentences before compound sentences). In some situations, however, the order in which you present information is not so clear cut. I've found that training needs often call for presenting information in a different order than how things are done in the "real world." For example, consider creating an exercise about a seven-step process in which step #3 includes four sub-steps (a, b, c, and d). In the real world the process might be performed as steps "1, 2, 3a, 3b, 3c, 3d, 4, 5, 6, 7." However, in a training setting, you may want to present these steps differently to help your audience understand each step and sub-step of the process. For instance, you might focus on the main process and skip the sub steps (1, 2, 3, 4, 5, 6, 7), and then go back later in a separate exercise to cover the sub-steps (3a, 3b, 3c, 3d).

In addition to developing your design documents, you might also need other types of design work at this stage. If your instructional materials are going to use illustrations or diagrams, you might get a graphic designer involved to start working on prototypes to identify the appropriate graphic design style for your materials. For instance, will you use rough, hand-drawn illustrations or clip art? If you are creating new materials from scratch, or don't have established templates and standards, you'll also want to design the templates and standards you'll use as you develop your materials. This can include presentation templates, as well as text document templates that specify fonts, heading sizes, and other formatting details. Using templates like this can help ensure consistency in your materials.

As we saw in Chapter 11, instructional design is an area where the Imagineering Process works on both the macro and micro levels (see "Scalability: Macro and Micro"). For example, the Design stage for large-scale projects such as a training program with several major courses might comprise a series of "micro" projects in which you go through an instance of the Imagineering Process for each course as you conceptualize, develop, design, model, and build the various courses that comprise your training program.

As you work through this stage, consider setting up some milestones, similar to the 30%, 60%, and 90% milestones the Imagineers use, to make sure your overall design effort is staying true to the project's creative intent. These are especially helpful if you have multiple people working on different aspects of the project, as they provide an opportunity for folks to look at each other's work and identify any potential conflicts or gaps.

Lastly, as you design your instructional materials in detail, use the principles of the Imagineering Pyramid to help make your materials as effective and engaging as possible. Are you:

- Using your subject matter to inform all decisions about your project? (It All Begins with a Story)
- Staying focused on your objective? (Creative Intent)
- Paying attention to every detail? (Attention to Detail)
- Using appropriate details to strengthen your story and support your creative intent? (Theming)
- Organizing your message to lead your audience from the general to the specific? (Long, Medium, and Close Shots)
- Attracting your audience's attention and capturing their interest? (Wienies)
- Making changes as smooth and seamless as possible? (Transitions)
- Focusing on the big picture? (Storyboards)

- Introducing and reinforcing your story to help your audience get and stay engaged? (Pre-Shows and Post-Shows)
- Using the illusion of size to help communicate your message? (Forced Perspective)
- Simplifying complex subjects? ("Read"-ability)
- Keeping the experience dynamic and active? (Kinetics)
- Using repetition and reinforcement to make your audience's experience and your message memorable? (The "it's a small world" Effect)
- Involving and engaging your audience? (Hidden Mickeys)
- Consistently asking, "How do I make this better?" (Plussing)

Your work in this stage should result in a set of design documents, including detailed outlines or mind maps, storyboards, and scripts that describe how to construct the materials for your project.

CONSTRUCTION

Build the actual project, based on the design developed in the previous stages.

In this stage you create the actual materials that will be used when teaching your audience. Instruction materials can include presentation slides, exercise packets, activity guides, sample data, job aids, recordings, or any number of other documents. The format(s) you chose for your project (in the Blue Sky and Concept Development stages) will play a big role in determining the specific materials you need to create.

For written materials, this means creating documents, writing text, creating illustrations or diagrams, and putting it all together in the proper format. A common approach to instructor-led training is to create presentation slides for lecture content, and an accompanying activity or practice guide that outlines hands-on activities in a formatted text document. Writing materials like this means lots of back and forth between the presentation slides and the text document to ensure they complement each other and flow together.

As you build these materials, you may find yourself re-thinking the order and sequence of your content. Even with the most detailed outline prepared from the most thoughtful examination of the subject matter, it's not uncommon to discover a better way to organize your material once you're in the middle of developing the content. This is because despite our best efforts, the outlines and documents we develop during the Concept Development and Design stages are in a sense predictions about the best way to present our information. It's often not until we get to actually developing the content that we figure out how best to put it all together. And even then we're predicting. After we present the content to our audience, we may find a need to refine things even further (as the saying goes, "No plan survives contact with the enemy"). Author Steward Brand provides a useful metaphor for this in his book *How Buildings Learn: What Happens After They're Built* when he writes, "All buildings are predictions. All predictions are wrong." So don't fret if you end up moving things around as you write your content. It's all part of the process.

For recorded materials, this stage is where you create your raw audio and video recordings, and assemble and edit them as needed. Some media tools allow recording of audio and video together, while others allow you to record them separately and edit them together. In my experience, the latter approach is easier in the long run, but in the end it's a matter or personal preference. If you're developing interactive e-learning content, this stage is where you create your modules using tools such as Adobe Captivate.

As you work on your materials, pay attention to formatting and the theming of your content. Theming is one of the foundational principles of the Imagineering Pyramid and plays a role in many types of creative projects. In instructional design, theming is all about making sure your content delivers its message in a clear and consistent manner. Sometimes inconsistent theming can distract your audience, and you don't want to do anything that's going to take your audience out of the experience. Theming in instructional materials includes not just using appropriate words or images, but also includes consistent use of templates and styles and fonts and colors. Consistent theming fades into the background, while inconsistency in theming can jar your audience and distract them from your content.

Lastly, as you build and develop your content, it's a good idea to use a form of "Test and Adjust", and test your practices and exercises as you work. Like the order and sequence of your content, your scripts are predictions and may need to be adjusted as you fully develop your content. This is especially true in situations in which exercises or practices build on each other, as is often the case when developing training for software applications; the first few exercises involve setting up data that is used in later lessons and exercises. If the later exercises don't work as expected, you may need to adjust the earlier exercises to make sure the data is set up correctly.

MODELS

Test and validate your design at each stage to help solve and/or prevent problems that may arise during the design and construction process. When developing instructional materials, models can take many forms and can be used in any/all of the stages of the Imagineering Process. Below are a few examples of the ways in which models can be used in instructional design.

Early outlines, storyboards, and mind maps can help you identify and solve potential challenges, particularly around the subject matter you plan to cover and the order in which to you plan to present your material.

Sharing early drafts of presentation slides with stakeholders and colleagues can help you work out a standard approach to structuring lessons

If you're developing media files such as recorded audio or video, creating short, simple, and unpolished prototypes and mockups is a good way to get stakeholder buy-in before investing the time and energy to script, record, edit, and produce full-length media.

If your courses, modules, or lessons contain hands-on practices, labs, or demonstrations, you will need to test them to make sure they work as planned. These tests are essentially models that can often lead to minor (and sometimes not-so-minor) adjustments in your lessons. For instance, when creating training materials for software applications, it's common to have connections and/or dependencies between the practices and labs (the data you create in practice #1 is used or referenced in practice #2 or 3, etc.). Creating tests and models of your practices can help you identify and plan for these types of dependencies or connections.

EPILOGUE: OPENINGS, EVALUATIONS, AND SHOW QUALITY STANDARDS

Present your project to your audience, allow them to experience it, and evaluate its success and effectiveness over time.

Openings in the world of instructional design are when we deliver the course to various audiences.

When developing training content, a common early delivery is a "train the trainer" session, in which the people who will one day deliver the training experience it as students as part of their preparation for eventually teaching the class on their own. "Train the Trainer" courses are often delivered by either subject matter experts or the instructional designers who developed the course. The goal of these sessions is to enable people to be able to deliver your training content on their own.

Another common type of early training delivery is what's know as a "first teach." This is the first time regular students or participants will experience the course. These are usually delivered either by instructors who have gone through some form of "train the trainer" course, or by subject matter experts or content developers.

After a class has been delivered, we want to solicit feedback from students and participants, typically via surveys and course evaluations, but it can come from instructors as well. However you obtain your feedback and evaluation data, use it to revise, correct, and "plus" your content so that future audiences will have an even better educational experience. In rare and extreme cases, feedback might reveal that the course either wasn't truly necessary or was the wrong way to address your original Need (this type of feedback is not common, but unfortunately is not unheard of either).

Feedback from initial deliveries ("train the trainer" and "first teach") is especially useful, since they often represent the first time the content will be shared with an audience that is not familiar with it, and thus can provide a fresh and outside perspective. Feedback from initial deliveries can often point to areas where you have either not enough or too much detail, or areas where you can refine your mix of different types of content (lecture content vs. hands-on practice vs. quizzes, etc.).

If your courses will be delivered on a regular basis to different audiences over the course of months and years, consider adopting a variant of WDI's Show Quality Standards to make sure that your materials continue to meet the original Need and Requirements, as well as your original creative intent. In the case of content whose subject matter may change, such as training materials for software applications that are enhanced over time, content revisions intended to update the material for changes in subject matter also provide a good opportunity for refreshing the content and revising and adjusting things. Remember that even the best materials can be plussed.

The Imagineering Process and ADDIE

Another way to look at applying the Imagineering Process to instructional design is to examine how it aligns with other instructional design process models. One specific process model commonly used in the training industry is known as ADDIE. Let's look briefly at how the Imagineering Process can work with and within the ADDIE model.

ADDIE is a framework of generic processes for instructional designers. It represents a descriptive guideline for building effective training and performance support tools in five phases:

- **ANALYSIS** clarifies the instructional problems and objectives, and identifies the learning environment and learner's existing knowledge and skills.
- **DESIGN** deals with learning objectives, assessment instruments, exercises, content, subject matter analysis, lesson planning, and media selection.
- **DEVELOPMENT** is when instructional designers and developers create and assemble content assets blueprinted in the design phase.
- **IMPLEMENTATION** develops procedures for training facilitators and learners. Training facilitators cover the course curriculum, learning outcomes, method of delivery, and testing procedures.

- **EVALUATION** consists of two aspects: formative and summative. Formative evaluation is present in each stage of the ADDIE process, while summative evaluation is conducted on finished instructional programs or products.

The following table shows how the ADDIE phases align with the Imagineering Process stages:

ADDIE	The Imagineering Process
Analysis	Prologue, Blue Sky
Design	Concept Development, Design, Models
Development	Construction, Models
Implementation	Construction, Models, Epilogue (Openings)
Evaluation	Epilogue (Evaluations and Show Quality Standards)

This comparison shows the similarity between the two models, and how they complement each other. Both models cover similar ground, but the Imagineering Process breaks the process down into more distinct pieces, and each ADDIE phase comprises one or more Imagineering stages.

There are a few ways in which combining the Imagineering Process with ADDIE can be effective:

First, the distinctions between Prologue and Blue Sky (identifying your Need, Requirements, and Constraints vs. coming up with a solution to the training problem) in the Analysis phase and between Concept Development and Design (what we need to build vs. how we will build it) in the Design phase are valuable. The types of questions and approaches you use for a needs analysis are different from those you use when brainstorming ideas or developing initial concepts. Likewise, the level of detail used when fleshing out your idea is different than that needed to create specifications and design documents (though one way to look at the ADDIE Design phase is as the merging of the Concept Development and Design stages). The Imagineering Process stages provide a means to break down the ADDIE phases into smaller and more distinct pieces.

Second, the basic ADDIE model doesn't include iteration. It's intended as a more linear model, in which you move from phase to phase without going back to revisit prior phases.

Employing the iterative nature of the Imagineering Process both within ADDIE phases (for example, iterating between Concept Development and Design in the ADDIE Design phase) and between phases (iterating between Blue Sky and Concept Development or between Design and Construction) can help ensure that as new requirements are discovered (forcing you to re-evaluate some of your decisions) or as new ideas are conceived (providing new options) you can employ them in your project.

Lastly, ADDIE's formative evaluation encourages you to evaluate your work and look for ways to improve or plus your project at each phase.

Imagineering Process Checklist Questions: Instructional Design

Stage	Questions
Prologue: Needs, Requirements, and Constraints	▪ What is your training course about? ▪ What are the learning objectives of your course or lesson? ▪ Who is your target audience? ▪ What type of constraints do you have to work within?
Blue Sky	▪ What does your audience need to know? ▪ What is your training project about? What is your subject matter? ▪ What is your creative intent with this course? ▪ What teaching/learning formats do you want to use? ▪ Do you have standard formats you use for your courses? ▪ What is the best way to convey to your audience what they need to know?
Concept Development	▪ Are you using questions to find aspects of your concept that are vague or undefined? ▪ Are you addressing all of the Requirements and Constraints that you identified earlier? ▪ Have you identified the number of courses you need to develop? ▪ Have you done enough research?

Design	<ul><li>Have you developed detailed outlines and storyboards that clearly set forth what needs to be built and how it should be built?</li><li>Do you need to write scripts for certain parts of your project?</li><li>Have you considered the Transitions within your courses, modules, and lessons to make sure you're presenting your material in the best order and sequence?</li><li>Are you looking for ways to leverage the principles of the Imagineering Pyramid to more effectively communicate your content?</li></ul>
Construction	<ul><li>Have you identified areas where you've had to do things differently than originally planned?</li><li>Are you paying attention to the "Theming" of your materials? Are you being consistent with fonts, colors, templates, etc.?</li><li>Are you using "Test and Adjust" as you develop your content to make sure that everything works as expected?</li></ul>
Models	<ul><li>Are you sharing outlines, storyboards, and mind maps with stakeholders to identify and solve potential challenges?</li><li>Are you creating prototypes or mockups of recorded content to get stakeholder buy-in?</li><li>Are you testing your labs or practices to ensure they work as expected?</li></ul>
Epilogue: Openings, Evaluations, and Show Quality Standards	<ul><li>How are you soliciting feedback on your instructional materials?</li><li>Are you looking for ways in which you can "plus" your content?</li><li>Can you use your own type of Show Quality Standards to keep your instructional content fresh and relevant?</li></ul>

CHAPTER FIFTEEN

Imagineering Leadership and Management

In previous chapters, we looked at the stages of Imagineering Process from the point of view of working on creative projects ourselves. From that perspective, we apply the process to our own projects and ideas where we're the ones doing most of the work. But there is likely to come a time for many of us (perhaps all of us) when we move from doing the work ourselves to leading or managing others doing the work on creative projects. When this happens, the manager or leader's job is to help facilitate the process and support their team as they take a project from concept to reality.

In this chapter, I want to take a slightly different approach and look at the role of a leader or manager of a team using the Imagineering Process to bring a creative idea to life. Some of you might be saying to yourselves, "I'm not involved in management or leadership." When we think of management and leadership we typically think of them in the context of business, but both apply outside of business as well. Just like I used a broad view of game design in Chapter 13, I'm taking a broad view of management and leadership here. Any time you need to coordinate or lead the efforts of people to accomplish a specific creative project, goal, or objective you're engaged in management and leadership. This can include something as simple as helping your children organize how they complete their homework or as involved as leading a community volunteer project involving dozens or hundreds of volunteers.

Before we go on, I should clarify that even though this chapter is focused on management and leadership, I don't mean to suggest that they are the same thing. They are far from it. In simple terms, leadership is about creating vision, while management is executing on that vision. In this chapter we're going to focus primarily on the latter: execution of a vision from Blue Sky to reality.

Leading Teams Through the Imagineering Process

In this section we'll look at the role of a leader or manager in each stage of the Imagineering Process, and how leaders can support their team along the way.

One thing to note up front is that in your role as leader or manager, you likely either represent the project's stakeholders or are one of the project stakeholders yourself. This means part of your job is to facilitate meetings between your team and the stakeholders, such as when the team has completed work at a specific stage and is ready to present it to stakeholders for approval to move on to the next stage (for example, between Blue Sky and Concept Development or between Concept Development and Design). It also means that your team will often need to turn to you for clarification and additional information about the project.

PROLOGUE: NEEDS, REQUIREMENTS, AND CONSTRAINTS

Define your overall objective, including what you can do, can't do, and must do when developing and building your project.

Your initial role during this stage is to present your team with the project's Need, Requirements, and Constraints. You might start with an initial briefing or meeting, but will most likely need to follow up over time with your team members to make sure they fully understand the Need driving the project. Remember in Chapter 4 when we talked about identifying the "real" Need behind your project and understanding the "why" behind that Need? Part of your job in leading and managing a team during this stage is to help them ask the right questions as they evaluate the project's Need, Requirements, and Constraints so they are fully enabled to move on to the next stage.

BLUE SKY

Create a vision with enough detail to be able to explain, present, and sell it to others.

An important role for managers and leaders in the Blue Sky stage is to organize, plan, and facilitate brainstorming sessions. Having a dedicated facilitator during brainstorming can help the other participants by freeing them up from having to worry about the time, or keeping everyone on topic, or enforcing the rules (no blocking means NO BLOCKING). Some ideas to help here include:

- When planning your brainstorming sessions, consider distributing a brief description of Wilson's "7 Agreements of Brainstorming" from *HATCH* prior to the meeting so that everyone is aware of the rules and how the session will be run.

- Before the meeting you should also spread the word among the team about the subject matter and the Need to be addressed (this is *HATCH* agreement #1) during your brainstorming sessions.

- You should also take responsibility for capturing and keeping all of the ideas generated during brainstorming sessions. If you and your team ever need to revisit this project, some of the ideas you came up with originally may prove useful. Remember, there is no such thing as a bad idea.

When the team moves on from brainstorming into some initial concept design work, you can help them along the way by asking questions to sharpen their ideas and help them stay on target. You can also remind them of any Requirements or Constraints that they haven't focused on yet (you'll do this again in the next stage).

Beyond facilitating and leading brainstorming sessions and supporting concept design, another important job for managers and leaders during this stage is to help the team define their story (subject matter) and creative intent. These are critical elements to any creative project and the team can't move forward to Concept Development until they have the basics of these two things worked out.

Lastly, you need to help the team when they present their initial concept to the project stakeholders. Remember the goal of this stage: *create a vision with enough detail to be able to explain, present, and sell it to others.* Perhaps the most important part of your job at this stage is to help the team stay focused on this goal.

CONCEPT DEVELOPMENT

Develop and flesh-out your vision with enough additional detail to explain what needs to be designed and built.

As your project moves into this stage, your role as leader and manager is to support the team as they flesh out and expand upon their initial concepts so they fully understand what they need to design and build. In Chapter 6 we saw that questions are a useful tool to learn what you don't yet know about your project. Help your team by asking questions that address the less defined aspects of their Blue Sky design. You should also help the team look for assumptions (especially unquestioned assumptions like the need for seats in the Apollo Lunar Module) and make sure they revisit any Requirements and Constraints that they haven't fully addressed yet.

Another key job during this stage is to keep the team focused on their story (subject matter) and creative intent. When we delve deep into creative work it's easy to get sidetracked from time to time, and part of your job as leader is to help the team stay focused. During this stage, as team members flesh out the project's initial concept, it's natural to come up with new ideas and concepts. As these new ideas are introduced, ask the team questions like the following to help them stay on track:

- Does this idea fit with our story and subject matter?
- Does this idea support our creative intent?

Like you did during Blue Sky, you need to also help your team present their fully developed concept to your project stakeholders before they can move on to the next stage. As part of the transition from Concept Development to Design, you should also evaluate the feasibility of the project based on factors such as cost, technology schedule, and others to make sure the project is achievable.

DESIGN

Develop the plans and documents that describe and explain how your vision will be brought to life.

When you lead a team into this stage, your job is again to support and facilitate the team as they develop the detailed design documents needed to bring your project to life.

At this stage you need to consider what different types of expertise are applicable to your project, and what specific disciplines need to be involved in its designing. If you don't have this expertise on your team, you need to be creative to fill in any specific skill or knowledge gaps.

In the likely event that your team members are going to split up the design work, a big part of your job is to make sure they don't go off entirely on their own and pay no attention to what others are doing. You need to encourage (and in some cases require) collaboration and communication between the team members to make sure their design work fits together and that the overall project hangs together.

A useful tool for this is to identify milestones (similar to the Imagineers' 30%, 60%, and 90% milestones) at which point team members working on different aspects of the design will review each other's work and identify any potential conflicts or gaps, and make sure the overall design effort is staying true to the project's creative intent.

If your project is made up of several smaller parts (and most are), you should identify the macro and micro levels of design, and help the team make sure that the work at the micro level supports the macro level, and vice versa.

CONSTRUCTION

Build the actual project, based on the design developed in the previous stages.

During this stage you must lead not only the construction of the individual components of your project, but also their assembly into the final product.

As your team works on building the various pieces of the project, you can help by addressing any issues and challenges that arise along the way. For example, if building training content for a software application, you can help by arranging for and providing a training environment the team can use when building out practices and hands-on exercises.

You should also consider how you can support the team as they test the project during this stage (similar to the Imagineers' Test and Adjust stage). For instance, you might serve as a test audience for the project, experiencing it like an outside audience will when the project is completed.

MODELS

Test and validate your design at each stage to help solve and/or prevent problems that may arise during the design and construction process.

As the team moves from stage to stage of the process, you should be reminding and encouraging them to validate and test their plans through models, prototypes, and mockups. You can also help test and evaluate the team's designs by reviewing the models and testing the prototypes they build.

EPILOGUE: OPENINGS, EVALUATIONS, AND SHOW QUALITY STANDARDS

Present your project to your audience, allow them to experience it, and evaluate its success and effectiveness over time.

When your team completes the Construction stage it's time for them to share their project with an audience and the stakeholders. In your role as leader you should facilitate this step.

When it comes to presenting the project, you can help your team by arranging to share it with its intended audience. Different types of projects can have different types of audiences, whether they be customers buying a product, participants experiencing an presentation or event, students participating in a training class, or whatever is appropriate. You might even want to volunteer to be the first audience. Experiencing the final product can be quite different than what you were party to during its development.

You should also work with the team to develop the criteria by which you can evaluate the success of the project. Basic criteria would include asking if the project meets its original Need and Requirements and if it violates any of the project's Constraints. As we saw in Chapter 10, you could also evaluate the project using the Imagineering Pyramid principles (see Chapter Ten,

Epilogue: Openings, Evaluations, and Show Quality Standards).

Hopefully over time different audiences will continue to experience your team's project, and eventually you will want to work with your team to develop something akin to Imagineering's Show Quality Standards to make sure that the project continues to serve its original objective. In some cases, circumstances and audiences change and a project that originally met its objective might no longer do so, and part of your job as a leader is to help your team identify when this happens, and coach them in how to address potential changes.

The Imagineering Process and Problem Solving

Most of the examples we've looked at in previous chapters are new creative "things" (for lack of a better word). Whether they be games, promotional logos, birthday party activities, or a spacecraft, the end result of most of these is a physical product of some sort. But in addition to helping you build new things, the Imagineering Process can also be used as a problem-solving process.

When in a management or leadership role, it's not uncommon to be presented with problems that need to be solved. When this happens to you (and it will), consider using the Imagineering Process as a model for developing a solution to the problem.

Let's look briefly at how the Imagineering Process works when used in problem-solving:

Prologue
Identify the problem. This is your Need, and will very often be presented to you by your management or other parties looking to you for help. Problems often have their own Requirements and Constraints (except for simple problems, but people wouldn't be coming to you if their problems were simple).

Blue Sky
Develop an idea to solve the problem. Use brainstorming or other ideation techniques to come up with various ideas and solutions to the problem, and select the best idea with which

to move forward. And be sure to keep the ideas you don't pursue—there is *no* such thing as a bad idea after all.

Concept Development

Flesh out the solution to make sure it addresses all aspects of the problem. This can involve considering the different needs of different stakeholders impacted by the problem. For example, in a corporate setting, if a problem involves a specific tool such as an HR application, your solution must meet the needs of the different organizations with the company that use that tool.

Design

Fully detail the solution so you're able to implement it. Work out the specific steps of your solution so that you can implement it appropriately.

Construction

Implement and/or put your solution into practice. This is where the process can differ a little from our previous examples.

- If your solution is a process or procedure, this where you have people begin to use the new process.
- If the solution includes a work product (a presentation or document) you would build those products at this stage, and share it with your stakeholders and audience in the Epilogue.

Models

If your solution involves documents or tools, you will want to experiment with prototypes and mockups during Blue Sky, Concept Development, Design, and Construction.

Epilogue

Share the results of your implementation with your original stakeholders and other interested parties and evaluate its effectiveness. If your solution is going to be an ongoing thing (such as a new process or policy), continually evaluate its effectiveness over time and adjust as needed.

Imagineering Process Checklist Questions: Leadership and Management

Stage	Questions
Prologue: Needs, Requirements, Constraints	▪ How can you help your team thoroughly evaluate your Need, Requirements, and Constraints? ▪ Does your team know what they really need to create?
Blue Sky	▪ Have you shared the "7 Agreements of Brainstorming" from *HATCH* with the team? ▪ Are you helping the team stay on target? ▪ How can you help the team define their story and creative intent?
Concept Development	▪ How can you help your team refine their ideas? ▪ Are you helping your team stay focused on your story and creative intent? ▪ What don't you and your team know about your project yet? ▪ Are you using questions from the Imagineering Pyramid to help expand upon the team's initial designs?
Design	▪ Are team members collaborating and communicating as they work on separate parts of the project? ▪ What are the milestones at which you should have team members check in with each other? ▪ Are there specific design disciplines needed for
Construction	▪ How can you help your team as they build the pieces and components of the project? ▪ Are you facilitating a "Test and Adjust" phase during your Construction stage?
Models	▪ Is your team creating models, prototypes, and mockups of their project? ▪ How can you help test your team's design?
Epilogue: Openings, Evaluations, and Show Quality Standards	▪ How are you evaluating the success of your project? ▪ What form of criteria is your team using to evaluate the project? ▪ Are you helping your team gather audience feedback from its audience? ▪ Does your team have its own form of Show Quality Standards?

Final Thoughts

In this chapter I want to revisit the basic premise of this book (outlined in the Introduction), present you with a challenge, and offer some parting thoughts.

Creativity and the Imagineering Process

"Creativity is not just for artists. It's for business people looking for a new way to close a sale; it's for engineers trying to solve a problem; it's for parents who want their children to see the world in more than one way." These words from renowned choreographer Twyla Tharp are at the heart of what this book is about: creativity is for everyone, and everyone can be creative. I truly believe that, and I hope you do as well.

So, what drives creativity? Ideas. Ideas are the sparks that lead to new inventions, new stories, and new...well, everything. Everything we experience in the world started out as an idea. But if we look closely at the role of ideas in the creative process, we find a paradox: ideas are at the same time the *most important* and the *least important* part of any creative project.

They are the *most* important part because every creative project starts with an idea. But at the same time (here comes the paradox), they are also the *least* important part. How can that be? As Guy Kawasaki puts it: "Ideas are easy. Implementation is hard."

Put another way, generating ideas is the easy part; it's the execution of those ideas that's difficult. The real work is in taking ideas and bringing them to life. Even the best ideas in the world can't execute themselves, and without someone to execute them, even the best ideas in the world have little chance of becoming a reality.

So, where can we look to help us bring our creative ideas to life? I think we need a model for the creative process—an example that we can look to for concepts and principles that can be applied across a variety of fields in bringing creative ideas from concept to reality.

Where can we find a model like this? In a single word, Disneyland. As Garner Holt and Bill Butler put it:

> Disneyland is still the ultimate expression of the creative arts: it *is* film, it *is* theater, it *is* fine art, it *is* architecture, it *is* history, it *is* music. Disneyland offers to us professionally (and to everyone who seeks it) a primer in bold imagination in nearly every genre imaginable.

To be more specific, I believe the design and development of Disney parks, the practice known as Imagineering, provides an ideal model for the creative process.

THE IMAGINEERING PROCESS

Disney Imagineers employ many different tools, techniques, and practices in their work. Among these is a process that can serve as a model of the creative process for us in other fields that lie "beyond the berm." This is the Imagineering Process, and it looks like this:

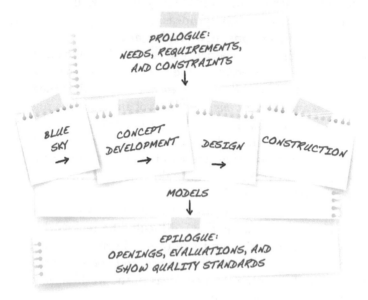

The stages of the Imagineering Process are:

Prologue: Needs, Requirements, and Constraints

Define your overall objective, including what you can do, can't do, and must do when developing and building your project.

Blue Sky

Create a vision with enough detail to be able to explain, present, and sell it to others.

Concept Development

Develop and flesh-out your vision with enough additional detail to explain what needs to be designed and built.

Design

Develop the plans and documents that describe and explain how your vision will be brought to life.

Construction

Build the actual project, based on the design developed in the previous stages.

Models

Test and validate your design at each stage to help solve and/or prevent problems that may arise during the design and construction processo.

Epilogue: Openings, Evaluations, and Show Quality Standards

Present your project to your audience, allow them to experience it, and evaluate its success and effectiveness over time.

My goal has been to describe these stages of the Imagineering Process, both how they are used by Walt Disney Imagineering, and how you can use them to bring *your* creative ideas to life.

It's my hope that this book has caused you to think about how you can apply the Imagineering Process to bringing your own creative ideas to reality.

A Challenge

So now I want to challenge you. I want you to use what you've read in this book and apply it to your own work. Think about a project you're currently working on (or will work on in the near future) and adopt the Imagineering Process to that project. As you work on that it and take your idea from concept to reality, think about how you can...

...define your overall objective, including what you can do, can't do, and must do when developing and building your project. (Prologue)

...create a vision with enough detail to be able to explain, present, and sell it to others. (Blue Sky)

...develop and flesh-out your vision with enough additional detail to explain what needs to be designed and built. (Concept Development)

...develop the plans and documents that describe and explain how your vision will be brought to life. (Design)

...build the actual project based on the design developed in the previous stages. (Construction)

...test and validate your design at each stage to help solve and/or prevent problems that may arise during the design and construction process. (Models)

...present your project to your audience, allow them to experience it, and evaluate its success and effectiveness over time. (Epilogue)

BUT MY PROJECT IS DIFFERENT

One thing you may find as you start to apply the Imagineering Process to your projects is that the manner in which you apply these stages will likely differ with different types of projects. For example, you probably noticed that the stages work slightly differently when applied to game design, instructional design, and management/leadership. It's also possible that one or more of the stages won't really apply to every creative project you work on. The truth is that not every tool is suited to every job.

If you find that one or more of the stages don't apply one of your projects, don't discard the rest of the process. Instead, take the time to consider those parts of the process that do apply and how you can use them in the development of your project. You might be surprised at just how flexible and versatile the Imagineering Process can be.

Thank You

In closing, I want to thank you for taking the time to read this book. Time is the one resource that we can't get more of, and I truly appreciate you deciding to spend some of your limited time with me to learn about Imagineering and the Imagineering Process. As I said in the Pre-Show, I'm still on my "journey into Imagineering" and this book has been the latest step for me on that journey. Writing this book forced me to look even deeper at the Imagineering Process and has only served to reinforce my passion for being a student of Imagineering, and now it's time for my next lessons.

Thank you for joining me on this part of my journey. I'm glad you came along.

My Imagineering Library

This appendix includes lists of the books, videos, and other resources in my "Imagineering Library."

Books

A Brush with Disney: An Artist's Journey, Told through the Words and Works of Herbert Dickens Ryman by David Mumford and Bruce Gordon

Building a Better Mouse: The Story of the Electronic Engineers Who Designed Epcot by Steve Alcorn & David Green

Design: Just For Fun by Bob Gurr

Designing Disney: Imagineering and the Art of the Show by John Hench (with Peggy Van Pelt)

Designing Disney: Imagineering and the Art of the Show (2nd Edition) by John Hench (with Peggy Van Pelt)

Designing Disney's Theme Parks: The Architecture of Reassurance by Karal Ann Marling (Editor)

Disneyland: Inside Story by Randy Bright

Disneyland: The Nickel Tour by Bruce Gordon and David Mumford

Disneyland Paris: From Sketch to Reality by Alain Littaye and Didier Ghez

Dream It! Do It!: My Half-Century Creating Disney's Magic Kingdoms by Martin Sklar

From Horizons to Space Mountain: The Life of a Disney Imagineer by George McGinnis

HATCH!: Brainstorming Secrets of a Theme Park Designer by C. McNair Wilson

It's Kind of a Cute Story by Rolly Crump

Jack of All Trades: Conversations with Disney Legend Ken Anderson by Paul F. Anderson

One Little Spark!: Mickey's Ten Commandments and The Road to Imagineering by Martin Sklar

Pirates of the Caribbean: From the Magic Kingdom to the Movies by Jason Surrell

Poster Art of the Disney Parks by Daniel Handke and Vanessa Hunt

The Art of Disneyland by Jeff Kurtti and Bruce Gordon

The Art of Walt Disney World Resort by Jeff Kurtti

The Disneyland Story: The Unofficial Guide to the Evolution of Walt Disney's Dream by Sam Gennawey

The Disney Mountains: Imagineering at Its Peak by Jason Surrell

The Haunted Mansion: From the Magic Kingdom to the Movies by Jason Surrell

The Haunted Mansion: Imagineering a Disney Classic by Jason Surrell

The Hidden Mickeys of Walt Disney World by Kevin and Susan Neary

The Imagineering Field Guide to Disney California Adventure at Disneyland Resort by Alex Wright

The Imagineering Field Guide to Disneyland by Alex Wright

The Imagineering Field Guide to Disney's Animal Kingdom at Walt Disney World by Alex Wright

The Imagineering Field Guide to Disney's Hollywood Studios at Walt Disney World by Alex Wright

The Imagineering Field Guide to Epcot at Walt Disney World by Alex Wright (Original)

The Imagineering Field Guide to Epcot at Walt Disney World: Revised and Updated by Alex Wright

The Imagineering Field Guide to the Magic Kingdom at Walt Disney World by Alex Wright (Original)

The Imagineering Field Guide to the Magic Kingdom at Walt Disney World: Revised and Updated by Alex Wright

The Imagineering Pyramid: Using Disney Theme Park Design Principles to Develop and Promote Your Creative Ideas by Louis J. Prosperi

The Imagineering Way: Ideas to Ignite Your Creativity by The Imagineers

The Imagineering Workout: Exercises to Shape Your Creative Muscles by The Disney Imagineers (Peggy Van Pelt, Editor)

The Hidden Mickeys of Walt Disney World by Kevin and Susan Neary

The Magic Kingdom Storybook by Jason Grandt

The Making of Disney's Animal Kingdom Theme Park by Melody Malmberg

Theme Park Design: Behind the Scenes with an Engineer by Steve Alcorn

Theme Park Design: The Art of Themed Entertainment by David Younger

Tony Baxter: First of the Second Generation of Walt Disney Imagineers by Tim O'Brien

Walt Disney Imagineering: A Behind the Dreams Look At Making the Magic Real by The Imagineers

Walt Disney Imagineering: A Behind the Dreams Look at Making More Magic Real by The Imagineers

Walt Disney's Epcot Center: Creating the New World of Tomorrow by Richard R. Beard

Walt Disney's First Lady of Imagineering, Harriet Burns by Pam Burns-Clair & Don Peri

Walt Disney's Legends of Imagineering and the Genesis of the Disney Theme Park by Jeff Kurtti

Periodicals

"Disneyland Is Good for You" (*New West*, December 1978). This interview by Charlie Hass with Imagineer John Hench is one of the earliest analyses of Disneyland and how the Imagineers create the magical experiences that guests enjoy when they visit Disney theme parks, and should be considered must-reading for anyone interested in learning about Imagineering. It is also one of the first articles (if not *the* first) to describe the idea of "reassurance" as one of the key hallmarks of Disney theme park design. In his foreword to *Designing Disney: Imagineering and the Art of the Show*, Imagineering executive Marty Sklar describes this interview as "the single clearest analysis of Walt

Disney's Magic Kingdom ever printed, before this book." The article is available online at Scribd.com (https://www.scribd.com/doc/17664805/Disneyland-Is-Good-For-You).

Disney twenty-three. This the magazine of D23, the official Disney fan club. Published quarterly, it features articles about Disney history, films and television shows, and theme parks. Articles are well-written and feature gorgeous photographs and illustrations.

The "E" Ticket. A magazine devoted to "collecting theme park memories" and named after the coveted "E ticket" at Disneyland Park that admitted the bearer to the most popular rides and attractions, *The "E" Ticket* was started by brothers Leon and Jack Janzen in 1986. Its intent was to not only provide a detailed history of the brothers' favorite park, but also give readers knowledge about the artists, Imagineers, and other creative individuals responsible for the magic of Disneyland.

In December 2009, Jack Janzen sold all of the assets of *The "E" Ticket* to the Walt Disney Family Museum, and several back issues, including 3 CD-ROM collections, are available from the Walt Disney Family Museum Store. *The "E" Ticket* is a fantastic resource for those interested in Imagineering and Disney theme park history.

Audio Books

Crump, Rolly. *More Cute Stories, Vol. 1: Disneyland History.*
———. *More Cute Stories, Vol. 2: Animators and Imagineers.*
———. *More Cute Stories, Vol. 3: Museum of the Weird.*
———. *More Cute Stories, Vol. 4: 1964/65 New York World's Fair.*
———. *More Cute Stories, Vol. 5: Animators and Imagineers Part 2.*
———. *More Cute Stories, Vol. 6: Knott's Bear-y Tales.*

DVDs

Bob Gurr: Turning Dreams Into Reality
Disneyland Resort: Imagineering the Magic
Magic Kingdom: Imagineering the Magic
The Science of Disney Imagineering: Design and Models
The Whimsical Imagineer

Walt Disney Treasures. Disneyland: Secrets, Stories & Magic
Walt Disney Treasures. Tomorrowland: Disney in Space and Beyond

Online and Other Resources

NASA Information Technology Summit Day 2—Walt Disney Imagineering: Jack Blitch. In 2010, NASA held its first information technology summit, and the speakers included Jack Blitch, vice president of Walt Disney Imagineering in Orlando. During his talk, Blitch described the process Walt Disney Imagineering uses when they design and build attractions, with a specific emphasis on their use of technology and business information modeling software. Video of the summit is available online at the C-Span website (http://www.c-span. org/video/?295077-1/nasa-information-technology-summit-day-2). The portion of the video featuring Jack Blitch starts at the 50:00-minute mark.

Creative Mornings. Walt Disney Imagineering: Jason Surrell, Alex Wright, & Jason Grandt; Walt Disney Imagineering: Wyatt Winter. Creative Mornings is a breakfast lecture series for the creative community held in various cities around the world. These two lectures in Orlando feature Imagineers from the Orlando branch of WDI. Video of the lectures are available online at the Creative Mornings website (http://creativemornings.com).

Dining with an Imagineer (Hollywood Brown Derby; Disney's Hollywood Studios). This is a dining experience available at Walt Disney World in which a small group of guests have lunch with an Imagineer. For details, visit https:// disneyworld.disney.go.com/dining/hollywood-studios/ dine-with-an-imagineer.

The Imagineering Model: Applying Disney Theme Park Design Principles to Instructional Design. As I mentioned in the Pre-Show, the Imagineering Process began as part of a presentation I gave at a learning and training conference in 2011 which I later updated for a webinar I presented in 2013. The *Imagineering Model* includes both the Imagineering Process and the Imagineering Pyramid. Both versions of the *Imagineering Model* presentation are available online at Scribd. com and SlideShare.net.

Imagineering Field Guide Indexes. As part of my research for this book (and my ongoing interest in Imagineering), I created indexes for all six of the Imagineering Field Guides, noting every reference to Imagineering terms and concepts, attraction names, Imagineers, and illustrations. These proved quite helpful when I was looking for examples of specific Imagineering principles while writing this book. I also combined these into a single master index of all the Imagineering Field Guides. You can find them on Scribd.com as well as in this public Dropbox folder: http://tinyurl.com/IFG-Indexes

Recommendations for Learning About Imagineering

I would recommend all of the books, magazines, DVDs, and videos listed above, and if you're a Disney park fan, I expect you may already own some (or all) of them. However, for those of you who might be newer to Imagineering, the list might seem daunting. With that in mind, here are a handful of specific recommendations for anyone interested in learning more about Imagineering:

Walt Disney Imagineering: A Behind the Dreams Look at Making the Magic Real. This beautiful coffee table book is one of the first published about Imagineering (originally published in 1996). It provides an excellent overview of Imagineering and the Imagineering process, and includes amazing concept artwork.

Designing Disney: Imagineering and the Art of the Show. This book by John Hench is, in the words of Imagineer Alex Wright on Twitter, "the closest thing we have to an Imagineering textbook." It explores the art and craft of Imagineering like few others, and should be in the library of anyone interested in understanding Imagineering and themed entertainment. It was a major resource for my research into Imagineering.

Imagineering Field Guides. Another primary source for this book were the Imagineering Field Guides by Imagineer Alex Wright. These small pocket guides provide concrete examples of Imagineering principles in practice. Don't be fooled by their small and simple appearance: they contain a wealth of information about Imagineering, as well as excellent photos and art work. I can't recommend them highly enough.

Walt Disney's Imagineering Legends and the Genesis of the Disney Theme Park. One of the best ways to study a subject is to learn about the people involved in that subject, and this book is a great way to put that into practice by learning about the earliest Imagineers. Written by Jeff Kurtti, it contains profiles of the first generation of Imagineers, starting with Walt Disney and the people he brought into WED Enterprises when first planning and designing Disneyland, and ending with John Hench (whom Kurtti calls the "Renaissance Imagineer"). This book reinforces the idea that Imagineering was born from people from different disciplines adapting the skills and techniques from their disciplines to craft the new art form of theme park design.

One Little Spark: Mickey's Ten Commandments and The Road to Imagineering. Published in September 2015, this book by Imagineering executive Marty Sklar explores "Mickey's Ten Commandments," which Sklar created "to explain to and remind Imagineers about the foundation principles on which our success has been built." For Sklar, "these principles have formed the standards the Imagineers have used to create Disney park experiences around the world." It also includes examples of how the Imagineers have both succeeded and failed at each of "Mickey's Ten Commandments" over the years, as well as a series of short essays by current (and former) Imagineers about how they joined Walt Disney Imagineering.

To Learn More...

If you're interested in learning more about the resources in my Imagineering Library, including links to online resources and videos, visit the "My Imagineering Library" board on Pinterest (https://www.pinterest.com/louprosperi/my-imagineering-library).

The Imagineering Pyramid

At various points throughout this book I've made references to the Imagineering Pyramid, the subject of the first book in the Imagineering Toolbox series. This appendix provides an overview of that concept.

The Imagineering Pyramid

The Imagineering Pyramid is a arrangement of fifteen important Imagineering principles, techniques, and practices used by Walt Disney Imagineering in the design and construction of Disney theme parks and attractions.

The principles in the Imagineering Pyramid each fall into one of five categories or groupings, each of which forms a tier within the pyramid.

The Tiers of the Imagineering Pyramid

Walt's Cardinal Rule

Making It Memorable

Visual Communication

Wayfinding

Foundations of Imagineering

TIER 1: FOUNDATIONS OF IMAGINEERING

The bottom tier of the pyramid includes the foundations, or "cornerstones," of Imagineering. These principles serve as the foundation upon which all other techniques and practices are based. There are five Imagineering foundations:

- **IT ALL BEGINS WITH A STORY**: using your subject matter to inform all decisions about your project
- **CREATIVE INTENT**: staying focused on your objective
- **ATTENTION TO DETAIL**: paying attention to every detail
- **THEMING**: using appropriate details to strengthen your story and support your creative intent
- **LONG, MEDIUM, CLOSE SHOTS**: organizing your message to lead your audience from the general to the specific

TIER 2: WAYFINDING

The second tier is focused on navigation and guiding and leading your audience, including how to grab their attention, how to lead them from one area to another, and how to lead them into and out of an attraction. The four Wayfinding principles are:

- **WIENIES**: attracting your audience's attention and capturing their interest
- **TRANSITIONS**: making changes as smooth and seamless as possible
- **STORYBOARDS**: focusing on the big picture
- **PRE-SHOWS AND POST-SHOWS**: introducing/reinforcing your story to help your audience get and stay engaged

TIER 3: VISUAL COMMUNICATION

The third tier includes techniques of visual communication that are used throughout the parks in different ways. You'll find examples of these in nearly every land and/or attraction. These principles are:

- **FORCED PERSPECTIVE**: using the illusion of size to help communicate your message
- **"READ"-ABILITY**: simplifying complex subjects
- **KINETICS**: keeping the experience dynamic and active

TIER 4: MAKING IT MEMORABLE

The fourth tier includes practices focused on reinforcing ideas and engaging the audience. It is the use of these techniques which helps make visits to Disney parks memorable. They are:

- **THE "IT'S A SMALL WORLD" EFFECT**: using repetition and reinforcement to make your audience's experience and your message memorable
- **HIDDEN MICKEYS**: involving and engaging your audience

TIER 5: WALT'S CARDINAL RULE

The top tier contains a single fundamental practice employed in all the other principles. I call this "Walt Disney's Cardinal Rule:"

- **PLUSSING**: consistently asking "How Do I Make This Better?"

The following diagram illustrates how these principles are arranged within the Imagineering Pyramid:

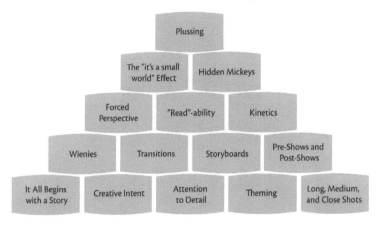

The Imagineering Pyramid

The Imagineering Process Checklist

Most of the previous chapters have included questions intended to help you apply the stages of the Imagineering Process to your own creative projects. This appendix collects all of those questions in a single Imagineering Process "checklist" that you can use to evaluate and "plus" your projects. It also includes a look at how the stages of the Imagineering Process intersect with the principles and practices of the Imagineering Pyramid.

Prologue: Needs, Requirements, and Constraints

Define your overall objective, including what you can do, can't do, and must do when developing and building your project.

General Questions	<ul><li>What is the problem I'm trying to solve? (What is the Need?)</li><li>What are the things I *must* do as part of this project? (What are my Requirements?)</li><li>What are the things I *can't* do as part of this project? (What are my Constraints?)</li><li>Have I thoroughly evaluated my Need, Requirements, and Constraints? Do I know what I *really* need to create?</li></ul>
Game Design Questions	<ul><li>What is your game about?</li><li>What type of game do you want or need to design?</li><li>What are the game play requirements?</li><li>What type of constraints do you have to work within?</li></ul>
Instructional Design Questions	<ul><li>What is your training course about?</li><li>What are the learning objectives of your course or lesson?</li><li>Who is your target audience?</li><li>What type of constraints do you have to work within?</li></ul>
Leadership / Management Questions	<ul><li>How can you help your team thoroughly evaluate your Need, Requirements, and Constraints?</li><li>Does your team know what they really need to create?</li></ul>

Blue Sky

Create a vision with enough detail to be able to explain, present, and sell it to others.

General Questions	• Are you using Wilson's 7 Agreements of Brainstorming? WDI's rules? Osborn's rules? • What other ideation techniques could you use if brainstorming isn't appropriate? • Do you have a vision for your project? • Have you developed your vision enough to be able to explain it to others? • Have you developed your idea enough to convince your stakeholders (whoever they may be) to proceed with your project?
Game Design Questions	• What is your game's story? • What is the experience you want players of your game to have? • Have you developed your vision enough to be able to explain it to others?
Instructional Design Questions	• What does your audience need to know? • What is your training project about? What is your subject matter? • What is your creative intent with this course? • What teaching / learning formats do you want to use? • Do you have standard formats you use for our courses? • What is the best way to convey to your audience what they need to know?
Leadership / Management Questions	• Have you shared the 7 Agreements of Brainstorming from HATCH with the team? • Are you helping the team stay on target? • How can you help the team define their Story and Creative Intent?

Concept Development

Develop and flesh-out your vision with enough additional detail to explain what needs to be designed and built.

General Questions	<ul><li>Have you developed your concept enough that you could turn it over to someone else and they could take the next steps?</li><li>Are you using questions to find aspects of your concept to further develop?</li><li>Is your concept feasible?</li><li>Have you done enough research?</li><li>Does your concept have implicit assumptions that could be questioned or challenged?</li></ul>
Game Design Questions	<ul><li>Have you defined Schell's basic elements of your game: Mechanics, Story, Aesthetics, and Technology?</li><li>Have you developed your game concept enough that you could turn it over to someone else and they could take the next steps?</li><li>Is your concept feasible?</li></ul>
Instructional Design Questions	<ul><li>Are you using questions to find aspects of your concept that are vague or undefined?</li><li>Are you addressing all of the Requirements and Constraints that you identified earlier?</li><li>Have you identified the number of courses you need to develop?</li><li>Have you done enough research?</li></ul>
Leadership / Management Questions	<ul><li>How can you help your team refine their ideas?</li><li>Are you helping your team stay focused on your story and creative intent?</li><li>What don't you and your team know about your project yet?</li><li>Are you using questions from the Imagineering Pyramid to help expand upon the team's initial designs?</li></ul>

Design

Develop the plans and documents that describe and explain how your vision will be brought to life.

General Questions	<ul><li>What types of design documents do you need to build your project?</li><li>What are the disciplines involved in designing your project?</li><li>Are there macro and micro levels of design for your project?</li><li>Are there parallels to Facility, Ride, and Show Design for your project?</li></ul>
Game Design Questions	<ul><li>What are the disciplines involved in designing your game?</li><li>What types of design documents do you need to build your game?</li><li>What specific types of design do you need to include in order to be able to implement your game?</li><li>Has your design work uncovered new Needs, Requirements, or Constraints that you hadn't identified previously?</li><li>Are there macro and micro levels of design for your game?</li></ul>
Instructional Design Questions	<ul><li>Have you developed detailed outlines and storyboards that clearly set forth what needs to be built and how it should be built?</li><li>Do you need to write scripts for certain parts of your project?</li><li>Have you considered the Transitions within your courses, modules, and lessons to make sure you're presenting your material in the best order and sequence?</li><li>Are you looking for ways to leverage the principles of the Imagineering Pyramid to more effectively communicate your content?</li></ul>

Leadership / Management Questions	<ul><li>Are you ensuring that team members collaborate and communicate as they work on separate parts of the project?</li><li>What are the milestones at which you should have team members check in with each other?</li><li>Are there specific design disciplines needed for your project?</li></ul>

Construction

Build the actual project based on the design developed in the previous stages.

General Questions	<ul><li>What are the specific tasks you need to complete to bring your idea to life?</li><li>What are the parts, pieces, and components of your project that can be built and tested prior to final assembly?</li><li>Are using your own version of Test and Adjust as you near completion of your project?</li></ul>
Game Design Questions	<ul><li>What are the tasks involved in assembling and constructing your game?</li><li>What sort of "Test and Adjust"effort are you using with your game?</li></ul>
Instructional Design Questions	<ul><li>Have you identified areas where you've had to do things differently than originally planned?</li><li>Are you paying attention to the "theming" of your materials? Are you being consistent with fonts, colors, templates, etc.?</li><li>Are you using "Test and Adjust" as you develop your content to make sure that everything works as expected?</li></ul>
Leadership / Management Questions	<ul><li>How can you help your team as they build the pieces and components of the project?</li><li>Are you facilitating a "Test and Adjust" phase during your Construction stage?</li></ul>

Models

Test and validate your design at each stage to help solve and/or prevent problems during the design and construction process.

General Questions	<ul><li>How can you use models at the various stages of the Imagineering Process?</li><li>Are you using playtesting to better understand how your audience will experience your project?</li><li>Are you using prototypes to refine your ideas?</li><li>Are there specific types of models you can use that are applicable to your field?</li><li>Could you use a form of workshopping to refine and develop your project?</li><li>Can you use storyboards to see the big picture of your project?</li></ul>
Game Design Questions	<ul><li>What types of models can you use as you design and build your game?</li><li>Are you playtesting your game? Are you playtesting it enough?</li><li>Are you creating prototypes and mockups of your game?</li></ul>
Instructional Design Questions	<ul><li>Are you sharing outlines, storyboards, and mind maps with stakeholders to identify and solve potential challenges?</li><li>Are you creating prototypes or mockups of recorded content to get stakeholder buy-in?</li><li>Are you testing your labs or practices to ensure they work as expected?</li></ul>
Leadership / Management Questions	<ul><li>Is your team creating models, prototypes, and mockups of their project?</li><li>How can you help test your team's design?</li></ul>

Epilogue: Openings, Evaluations, and Show Quality Standards

Present your project to your audience, allow them to experience it, and evaluate its success and effectiveness over time.

General Questions	• How will you share your project with your audience?
	• Can you use previews or soft openings to introduce your project to select audiences?
	• How are you going to evaluate the success of your project?
	• Do you have an equivalent of Show Quality Standards?
Game Design Questions	• Are you able to offer a preview release of your game to a select audience (such as crowdfunding backers)?
	• How will you handle errors and issues that are uncovered after the game has been released?
Instructional Design Questions	• How are you soliciting feedback on your instructional materials?
	• Are you looking for ways in which you can "plus" your content?
	• Can you use your own type of Show Quality Standards to keep your instructional content fresh and relevant?
Leadership/ Management Questions	• How are you evaluating the success of your project?
	• What form of criteria is your team using to evaluate the project?
	• Are you helping your team gather feedback from its audience?
	• Does your team have its own form of Show Quality Standards?

The Process and the Pyramid

This section looks at how the techniques from the Imagineering Pyramid and the stages of the Imagineering Process intersect.

Let me start with a caveat. As we look at the techniques and the process, one could argue that every technique applies to every step of the process. While that may be true, my goal here is to highlight the strongest correlations between the process and the principles.

For example, before you can move on from Blue Sky to Concept Development, you really need to know what your story (subject matter) is, you need to know your creative intent, and you want to look at how you can employ details to support both story and intent through theming, and how are you going to present your information to your audience through long, medium, and close shots.

When you get into Concept Development and Design, you're going to want to employ all (or nearly all) of the techniques we've looked at. The same applies to Construction as well (since that's when you physically implement all of your ideas).

	Blue Sky	Concept Development	Design	Construction	Models
It All Begins with a Story	X	X	X	X	X
Creative Intent	X	X	X	X	X
Attention to Detail	X	X	X	X	X
Theming	X	X	X	X	X
Long, Medium, and Close Shots	X	X	X	X	
Wienies		X	X	X	X
Transitions		X	X	X	X
Storyboards		X	X	X	
Pre-Shows and Post-Shows		X	X	X	X
Forced Perspective			X	X	X
"Read"-ability		X	X	X	X
Kinetics		X	X	X	X
The "it's a small world" Effect		X	X	X	X
Hidden Mickeys		X	X	X	X
Plussing	X	X	X	X	X

Bibliography

Books

Brand, Stewart. *How Buildings Learn: What Happens After They're Built*. New York. Viking. 1994.

Brown, Timothy. *Change by Design: How Design Thinking Transforms Organizations and Inspires Innovation*. New York. HarperCollins Publishers. 2009.

Cabrera, Alfredo with Matthew Frederick. *101 Things I Learned in Fashion School*. New York. Hachette Book Group. 2010.

Crump, Rolly. *More Cute Stories, Volume 5: Animators and Imagineers, Part 2*. Baltimore: Bamboo Forest, 2016.

Florida, Richard. *The Rise of the Creative Class, Revisited*. New York: Basic Books, 2012.

Frederick, Matthew. *101 Things I Learned in Architecture School*. Cambridge: The MIT Press, 2007.

Gennawey, Sam. *Walt Disney and the Promise of Progress City*. Winter Garden: Theme Park Press, 2014.

Hahn, Don. *The Alchemy of Animation: Making an Animated Film in the Modern Age*. New York: Disney Editions, 2008.

Hench, John, and Peggy Van Pelt. *Designing Disney: Imagineering and the Art of the Show*. New York: Disney Editions, Inc., 1998.

Imagineers, The, and Kevin Rafferty. *Walt Disney Imagineering: A Behind the Dreams Look at Making the Magic Real*. New York: Hyperion, 1996.

Imagineers, The, and Melody Malmberg. *Walt Disney Imagineering: A Behind the Dreams Look at Making More Magic Real*. New York: Disney Editions, Inc., 2010.

Kelly, Thomas J. *Moon Lander: How We Developed the Apollo Lunar Module*. Washington: Smithsonian Institution Press, 2001.

Kelley, Tom and David Kelley. *Creative Confidence: Unleashing the Creative Potential Within Us All*. New York: Crown Business, 2013.

Korkis, Jim. *Who's the Leader of the Club?: Walt Disney's Leadership Lessons*. Winter Garden: Theme Park Press, 2014.

Kurti, Jeff. *Walt Disney's Imagineering Legends and the Genesis of the Disney Theme Park*. New York: Disney Editions, Inc., 2008.

Martin, Vibeke Norgaard with Matthew Frederick. *101 Things I Learned in Law School*. New York. Hachette Book Group. 2013.

Moran, Christian. *Great Big Beautiful Tomorrow: Walt Disney and Technology*. Winter Garden: Theme Park Press, 2015.

Prosperi, Louis J. *The Imagineering Pyramid: Using Disney Theme Park Design Principles to Develop and Promote Your Creative Ideas*. Winter Garden: Theme Park Press, 2016.

Schell, Jesse. *The Art of Game Design: A Book of Lenses*. Burlington: Morgan Kaufmann Publishers, 2008.

Sims, Peter. *Little Bets: How Breakthrough Ideas Emerge From Small Discoveries*. New York: Free Press, 2011.

Sklar, Marty. *Dream It! Do It! My Half-Century Creating Disney's Magic Kingdoms*. New York: Disney Editions, Inc., 2013.

Sklar, Marty. *One Little Spark: Mickey's Ten Commandments and The Road to Imagineering*. New York: Disney Editions, Inc., 2015.

Surrell, Jason. *The Disney Mountains: Imagineering at Its Peak. New York:* Disney Editions, Inc., 2007.

Tharp, Twyla. *The Creative Habit: Learn It and Use It for Life*. New York: Simon & Schuster Paperbacks, 2003.

Williams, Pat, and Jim Denney. *How To Be Like Walt: Capturing the Disney Magic Every Day of Your Life*. Deerfield Beach: Health Communications, Inc., 2004.

Wilson, C. McNair. *HATCH: Brainstorming Secrets of a Theme Park Designer*. Colorado Springs. Book Villages. 2012.

Wright, Alex. *The Imagineering Field Guide to the Magic Kingdom at Walt Disney World*. New York: Disney Editions, Inc., 2005.

Wright, Alex. *The Imagineering Field Guide to Epcot at Walt Disney World*. New York: Disney Editions, Inc., 2006.

Wright, Alex. *The Imagineering Field Guide to Disney's Animal Kingdom Theme Park at Walt Disney World*. New York: Disney Editions, Inc., 2007.

Wright, Alex. *The Imagineering Field Guide to Disneyland*. New York: Disney Editions, Inc., 2008.

Wright, Alex. *The Imagineering Field Guide to Disney's Hollywood Studios at Walt Disney World*. New York: Disney Editions, Inc., 2010.

Wright, Alex. *The Imagineering Field Guide to Disney California Adventure at Disneyland Resort*. New York: Disney Editions, Inc., 2014.

DVDs
The Science of Disney Imagineering: Designs and Models. Disney Educational Productions. 2009. DVD.

Internet Sources
In addition to the books and DVDs listed above, I referenced these online sources while researching and writing this book:

Online Magazine Articles

Burkus, David. "Brainstorming is Dead; Long Live Brainstorming," *Forbes* (September 10, 2013). http://www.forbes.com/sites/davidburkus/2013/09/10/brainstorming-is-dead-long-live-brainstorming

Childs, William. "Feed Your Mind with These Must-Reads," *The Morning Call* (January 6, 2017). http://www.mcall.com/business/tech/mc-creativity-works-books-20170106-story.html

Childs, William. "The New Age of Innovation," *The Morning Call* (January 3, 2017). http://www.mcall.com/business/tech/mc-bill-childs-creative-age-20170103-story.html

Christensen, Clayton M, Scott Cook, and Taddy Hall. "What Customers Want from Your Products," *Working Knowledge* (January 16, 2006). http://hbswk.hbs.edu/item/what-customers-want-from-your-products

Kawasaki, Guy. "Ideas Are Easy, Implementation Is Hard," *Forbes* (November 4, 2004). http://www.forbes.com/2004/11/04/cx_gk_1104artofthestart.html

Mayer, Melissa. "Creativity Loves Constraints," *Bloomberg* (February 2, 2006). http://www.bloomberg.com/bw/stories/2006-02-12/creativity-loves-constraints

Rothman, Joshua. "Creativity Creep," *The New Yorker* (September 2, 2014). http://www.newyorker.com/books/joshua-rothman/creativity-creep

Blog Posts and Social Media

Hill, Jim. "Shanghai Disneyland Concept Art Reveals How Good the Imagineers Are at Hiding Things," *Jim Hill Media* (February 17, 2011). http://jimhillmedia.com/editor_in_chief1/b/jim_hill/archive/2011/02/17/shanghai-disneyland-concept-artreveals-how-good-the-imagineers-are-at-hiding-things.aspx

O'Keefe, Matt. "8 Key Principles That Disney Imagineers Use to Develop New Attractions," *Theme Park Tourist* (September 22, 2014). http://www.themeparktourist.com/features/20140922/29295/8-principles-disney-imagineer

Rohde, Joe. Instagram (March 28, 2017). https://www.instagram.com/p/BSKzDa_Ae5o

"What's in a Name?" *Re-Imagineering* (May 18, 2006). http://imagineerebirth.blogspot.com/2006/05/whats-in-name.html

Wright, Alex. Twitter (February 23, 2014). https://twitter.com/27000Acres/status/437806081259999232

Online Videos

Great Big Beautiful Tomorrow: The Futurism of Walt Disney. http://www.christianmoran.com/watch-the-films/feature-docs/great-big-beautiful-tomorrow-the-futurism-of-walt-disney

Websites

"18 Best Idea Generation Techniques." https://www.cleverism.com/18-best-idea-generation-techniques

"About Autodesk." https://www.autodesk.com/company

"Answers from a Spec Writer: Specifications." http://www.specificationsdenver.com

"Answers from a Spec Writer: For Non-Architects." http://www.specificationsdenver.com/for-non-architects.html

"BIM and the Future of AEC: What is BIM?" https://www.autodesk.com/solutions/bim

"Creative Mornings—Walt Disney Imagineering: Jason Surrell, Alex Wright, and Jason Grandt." https://creativemornings.com/talks/disney-imagineering/1

"Crush 'n' Gusher." https://disneyworld.disney.go.com/attractions/typhoon-lagoon/crush-and-gusher

"Lunar Module LM 2." https://airandspace.si.edu/collection-objects/lunar-module-2-apollo

"Microlearning for the Workplace." https://learningrebels.com/workshops/microlearning-for-the-workplace

"SCAMPER: Improving Products and Services." https://www.mindtools.com/pages/article/newCT_02.htm

"The 'E' Ticket" Now Available Online." http://www.waltdisney.org/blog/e-ticket-now-available-online

"The Science of Disney Imagineering: about the Series." http://dep.disney.go.com/imagineering.html

"The Untold Story of the Witches of Oz." http://wickedthemusical.com/the-show

"What Games Are: Waterfall Development." http://www.whatgamesare.com/waterfall-development.html

Acknowledgments

I have many people to thank who helped me make this book a reality:

- Readers of my first book who shared their reactions to it, and asked when the next book would be published. Their interest in my ideas about creativity and Imagineering were a great boost to my morale while I worked on this book.

- Family and friends who asked "How's your book going?" or "I heard you're writing a second book" and who gave me continued support, encouragement, and enthusiasm.

- John Fox at the Society of Applied Learning Technology, for giving me the first opportunity to share these ideas with an audience at the S.A.L.T. Conference in Orlando in February 2011.

- Bob McLain at Theme Park Pres,s for giving me the chance to share the Imagineering Toolbox with a larger audience through *The Imagineering Pyramid* and this book.

- Paul Brown at Laban Brown Design, for the Imagineering Process diagrams used throughout this book.

- The Themed Entertainment Association (TEA), for the use of the Project Development Process diagram.

- Jeff Barnes, David Burkus, Lee Cockerell, Brian Collins, Sam Gennawey, Jeff Laubenstein, Louis Lemoine, Matthew E. May, Daniel Pink, Josh Shipley, Jason Surrell, Jeff Tidball, Terry Weaver, and C. McNair Wilson, for reading an early draft of this book and providing helpful comments, feedback, and endorsements.

- Imagineer Jason Grandt, for the interesting and engaging stories he shared with my family and I during our "Dine with an Imagineer" in August 2010, for talking about Imagineering with me over dinner at Ragland Road in February 2011, and for putting up with all of my questions.

- Imagineer Alex Wright, for answering questions about Imagineering, the Imagineering process, and the Imagineering Field Guide series.
- Lastly, everyone with whom I have ever visited a Disney theme park, including family, friends, and the faculty, students, and chaperones of the 2012–2013 and 2016–2017 Wakefield High School Music Department. You help me see the Disney parks through new eyes each time I visit.

About the Author

Lou Prosperi worked in the game industry for 10 years as a freelance game designer and writer, and product line developer at FASA Corporation, where he worked on the *Earthdawn* roleplaying game. After leaving FASA, Lou went to work as a technical writer and instructional designer and has been in that role for the last 20 years, providing user and technical documentation and training for enterprise software applications. He currently manages a small team of technical writers and curriculum developers for a small business unit of a large enterprise software company.

Lou has been interested (or obsessed depending on who you ask) in Disney parks since his first visit to Walt Disney World on his honeymoon in 1993. A self-described "student of Imagineering," Lou has been collecting books about the Disney company, Disney parks, and Imagineering for the last 12+ years. He rarely passes up an opportunity to add new books to his Disney and Imagineering libraries, and is nearly always thinking about his next trip to Walt Disney World. Lou lives in Wakefield, Massachusetts, with his wife and children.

You can contact Lou via email or on social media at the following:

- LJP1963@aol.com
- http://www.facebook.com/lou.prosperi
- http://www.twitter.com/louprosperi
- http://www.pinterest.com/louprosperi
- https://roosevelt.academia.edu/LouProsperi

ABOUT THEME PARK PRESS

Theme Park Press publishes books primarily about the Disney company, its history, culture, films, animation, and theme parks, as well as theme parks in general.

Our authors include noted historians, animators, Imagineers, and experts in the theme park industry.

We also publish many books by first-time authors, with topics ranging from fiction to theme park guides.

And we're always looking for new talent. If you'd like to write for us, or if you're interested in the many other titles in our catalog, please visit:

www.ThemeParkPress.com

• •

Theme Park Press Newsletter

Subscribe to our free email newsletter and enjoy:

- ◆ Free book downloads and giveaways
- ◆ Access to excerpts from our many books
- ◆ Announcements of forthcoming releases
- ◆ Exclusive additional content and chapters
- ◆ And more good stuff available nowhere else

To subscribe, visit www.ThemeParkPress.com, or send email to newsletter@themeparkpress.com.

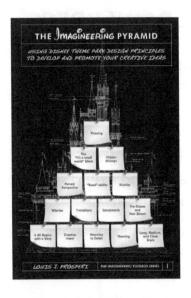

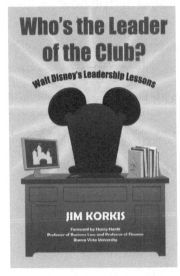

Read more about these books
and our many other titles at:

www.ThemeParkPress.com

15369547R00146